ENCOUNTERING THE ENVIRONMENT

ENCOUNTERING THE ENVIRONMENT

Articles from *Natural History*,
the Journal of The American Museum of Natural History

Edited by
Alfred Meyer

Van Nostrand Reinhold Company

NEW YORK / CINCINNATI / TORONTO / LONDON / MELBOURNE

Van Nostrand Reinhold Company Regional Offices:
New York Cincinnati Chicago Millbrae Dallas
Van Nostrand Reinhold Company International Offices:
London Toronto Melbourne

Library of Congress Catalog Card Number: 76-138488

Manufactured in the United States of America
Published by Van Nostrand Reinhold Company
450 West 33rd Street, New York, N.Y. 10001
Published simultaneously in Canada by Van Nostrand Reinhold Ltd.

15 14 13 12 11 10 9 8 7 6 5 4 3 2 1

PREFACE

Numerous prophets have put an ear to the railroad track and heard the approaching rumble of the Environmental Disaster Express. We all recognize the train; we know it is real and we know it can bring serious consequences to all those things on the earth that metabolize, or those that, in the scheme of biological succession, are slated to metabolize.

It is in the context of this rising body of evidence pointing to the possibility of extinction for many earth species, including man, that many scientists brood, for the present has become increasingly inescapable, increasingly threatening. Men of science have always known that a pernicious government could parasitize and even pulverize their science, but in a way that is accidental, like an element of the personal. The neat biological property called diversity, amazingly enough, also works in human societies. Thus far, western science has blossomed partially because it managed to thrive in certain places while it was being perverted in others. What no one was quite prepared to think about until recently was the possibility of extinction. Extinction in all species is global, permanent, and, like death, irreversible. No longer is the question of continued human intellectual aspiration a personal one or a governmental one. Now it must be evaluated in terms of the diminishing viability of the human species.

One still gets a sense of the placid from a visit to the American Museum of Natural History, but the sense is superficial, for the wheels are turning: we are all encountering the environment. It was in the middle of 1965 that those of us on the magazine staff began to agonize over our present roles and particularly, those in the near future. We attempted to bring to bear in our pages in a concerted way the specifics not only of environmental processes in a biological sense—which *Natural History* had been doing superbly for the sixty-five years it has been published—but also of how those processes were affected and shaped by social forces, whether political, commercial, or more fatuously historical. The vibrations in our offices when the first of the essays brought together in this volume appeared were positively clamorous. We shared the zeal of the innocent in thinking that all the world needed was enthusiasm, soul, and a little on the ball. However, like the March of Dimes before the Salk vaccine, we were unprepared for what turned out to be a revolution in our own bailiwick.

As I write, Earth Day has come and gone and the environmental din has subsided somewhat although, since it is approaching November, the speeches of candidates invariably include "The Environment" as a central issue, along with "Law and Order" and "The War." In certain areas, an environmental ethic has begun to seep into legislation. DDT has been banned. Doubts about the wisdom of assessing the goals of this nation in terms of growth or GNP are more commonplace. Furriers are nervous, as are automobile manufacturers. The SST is in serious disrepute; so is the Army Corps of Engineers.

The essays that comprise *Encountering the Environment* are each a separate encounter with one or even several aspects of the conditions that have precipitated these developments. They represent not only elements of an aroused scientific community, but of an aroused citizenry as well. If there is a single theme running through them it is something akin to the ambience that permeates the American Museum of Natural History. The placid quality mentioned earlier is indeed a function of a somewhat remote perspective. But there is something else at the Museum that underlies this necessary rhetoric of time and distance that is also central to the ambience. It is an unshakable belief in the integrity of life systems and an unspoken celebration of life itself.

Alfred Meyer
The American Museum of Natural History
Editor, Natural History

New York, N.Y.
November, 1970

CONTENTS

ENCOUNTERING THE ENVIRONMENT

1

"The New Conservation"

by Richard L. Means

A three-year study of citizen conservation groups led to Richard L. Means's sociological analysis of the "new conservation." The author of Time and Society: The Sociological Dimensions of History, *published in 1970, Dr. Means is a professor of sociology and anthropology at Kalamazoo College. A visiting lecturer at the Yale School of Forestry during 1968–69, Dr. Means continued to study community attitudes toward the environment—the theme of much of his writing, including* The Ethical Imperative: The Crisis in American Values, *published in 1969.*

The magic word for this decade is "relevance." Everybody seems to be caught up in the search for relevance. Ministers, students, psychiatrists, and even politicians eagerly respond to the endless incantation of this magic word. Fortunately, the question of "relevance to what?" has a way of intruding itself on the more somnolent forms of contemporary escapism. One cannot completely escape—unless in lunacy—the realities of one's own society and history. Thus there are objective realities, outside the framework of egotistical whimsy, to which man has to relate to find relevance and to test his own experience. The world of nature, the environment of man, may be such an anchor point where man can discover a partial sense of permanence in a world of shifting relativities.

There is evidence that many people are attempting, consciously or unconsciously, to speak to this issue—the problems of order and chaos. It seems that some of the best answers are coming from a group I call "the new conservationists."

It is true that different groups seem to march to different drums. Each one—black or white, Catholic or Protestant, rich or poor—puts forward a special interpretation of the world. Ideology, the preaching of group-interest, seems to be the ruling passion of the day. However, the question is seldom asked of how men and institutions may be related to the general welfare. Undoubtedly many people in our society have special social needs: the black ghetto mother, the chronically ill retiree, the psychotic, the drug addict, etc. But how may we transcend individual and private interests and speak intelligently and objectively about the general public good? Is there any place today—other than in the world of abstraction—where we can find hope, at least in part, that the public good will be represented? Again I believe there is an answer, and that answer lies in a growing social movement, "the new conservation."

What is this social movement? First, and paradoxically, there is a sense in which it is not new. Like every valid social movement in a democracy, its meaning is expressed in the lives of ordinary men and women. The "movement" is an irate housewife who tries to sue her state government after an airplane swoops down over her clean wash fluttering in the afternoon breeze and sprays it with a deadly, sticky insecticide. The movement is also a small group of men and women standing, during the Christmas season, beside a river in Connecticut. They hold their noses because of the stench and decay of the once clean river and pray to God for forgiveness for what man may do to the earth. It is, as well, scientists and citizens who meet before a state commission to protest a new highway threatening to destroy thousands of acres of marshland.

The truth is that the people involved in the new conservation come from all walks of life.

Second, most of the new conservationists are not sentimentalists or what are sometimes called "pure preservationists." They are often the sons and daughters of the corporate world who do not respond easily to romantic primitivism. They understand William H. Whyte's dictum about preserving land and open space (*The Last Landscape,* 1968) when he says "use it or lose it." And they understand why, when Steven T. Mather was made the first chief of the National Park Service, he insisted on building roads so the public could see the national parks. Although he was immediately accused of destroying the parks—some people wished to keep nature locked up as a private preserve, to be enjoyed only by those with a great deal of money or free time—it may prove to have been one of the great statesman-like acts of American history that this pure preservationist doctrine never prevailed. By and large, the new conservationists are, whether they know it or not, supporters of Mather's idea.

The general policy of the National Park Service and the Forest

Service has been—as much as their limited budgets allow—to open the public domain to camping, fishing, hiking, and other outdoor activities. As a result, millions of Americans living in cities have had the opportunity to secure at least a glimpse of the wonders of nature in America. Thus, many have been encouraged to increase their appreciation of natural beauty.

This, then, brings us to the third characteristic of the new conservation—that more and more Americans are not just limiting their perception of nature to what might be called "the great outdoors," but are applying (at an increasing rate) their moral and esthetic appreciations of nature to the total environment. It is a central characteristic of the new conservation also to be concerned with the complete environment —city and country. In the realm of nature, the old dichotomies between city and rural life no longer prevail in quite the same way. The mighty Hudson River is just as polluted miles above New York City as it is in the city itself—equally reprehensible in either instance.

A fourth aspect of the new conservation is its scientific character, especially its relationship to the relatively new science of ecology. (Ecology is the systematic study of the entire environment, or ecosystem, which includes the relations between land, plants, animals, *and* man.) The new new conservationists are not simply democratic, practical, and concerned with the total environment—they are also amazingly well versed in scientific matters related to the environment. If they lack expertise, more and more citizens' groups are co-opting environmental scientists into political action as counselors and guides (as in the case of the West Michigan Environmental Information Center). This brings us to a fifth characteristic of the movement—its stress on problem solving.

Since it is the essence of the new conservationists to search for scientific and rational descriptions of the causes of environmental contamination, and since each environmental problem is always part of an ecological system that is unique, the movement is partially saved from becoming politically doctrinaire. It cannot place the blame simply on one individual or class. In short, scapegoating is out.

Take the case of the pollution of Lake Erie. The runoff of phosphate fertilizers from hundreds of Ohio farms has had a significant role in the pollution of the lake. The phosphates (and nitrates) became nutrients for a fantastic growth of algae in the lake—accelerating the natural process called *eutrophication*. The western basin of Lake Erie is covered, during the summer, with an 800-square-mile mass of algae, two feet thick. This tremendous algal growth, in turn, removes much of the oxygen from the water, which makes life impossible for the more desirable forms of life (game fish, etc.), and also makes swimming a messy and risky business. As a result, the value of Lake Erie as a recreational resource has declined tragically.

But can one blame just the Ohio farmer for what has happened? Not exactly. There are many factors involved that mean the responsibility (if it can be established) has to be shared. Industries and municipalities that dump their wastes into the lake are also to blame. The truth is that no one could foresee the tragic consequences of the many oversights of the past. Also, certainly in the case of the farmer, competing values have been involved—the desire for high crop yields, an immediate gain, versus the results of fertilizer runoffs perhaps hundreds of miles away. It would be foolish to expect the typical farmer to be at all conscious of the consequences of his use of the new chemical fertilizers. Furthermore, more obvious targets responsible for the biological death of the lake have been readily at hand and thus are more salient targets for public anger.

For example, the flotsam of Ohio industrial life floating for these many years down the Cuyahoga River at Cleveland has been apparent to anyone with eyes to see. The river is an open sewer, and moved George Stewart, the nature writer, to refer to American civilization as a "nation of crud." The Cuyahoga River is perhaps the only river in the world that is a fire hazard; it is so full of oil and various products, all interacting, that it continually produces methane gas. In November of 1952 it caught fire and caused a million dollars' worth of damage—and it is reported that it is as much a hazard today as ever.

The new conservationists believe that those who directly pollute the environment should be policed; existing laws and penalties should be strengthened and enforced. Nevertheless, they recognize that pointing the finger of blame, in relation to past actions and in a categorical way, is not an adequate conservation policy. What is evident to the new conservationists is that the problem of environmental deterioration involves myriad factors and causes; and, in turn, it must be attacked on a broad front. The problem is, what do we do now?

From the scientific view, ecology provides part of the answer, for it describes the many interacting causes that have upset the traditional balance of nature, and it suggests that man must work to redress the balance or he is in trouble both from the point of view of public health and in terms of creating a livable environment.

As a consequence, an important characteristic of new conservation is that it is more than a practical, democratic, and scientific response of various citizens to America's eroding landscape. Its chief distinction (and in the long run its great potential contribution) is that it is giving birth to a viable doctrine of public welfare. Although nascent at this stage and just beginning to find its articulate spokesmen in such men as Barry Commoner (*Science and Survival,* 1967) and Paul Shepard (*Man in the Landscape,* 1967), there is evidently a groping for a new social ethic capable of relating man to nature—a search and desire for what Aldo Leopold years ago called an "ecological conscience."

This new ethic stresses the search for an optimum use of the realities of existence—air, water, land—as the very substratum of human life itself. The mode of consideration is objective and scientific (ecological). The idea is that men's actions must test themselves against what we do to nature, because in the long run our actions are transferred through nature to future generations.

The question now arises—if there is a new conservation movement (inchoate and just beginning), why did it arise at this stage of American history?

There are many plausible reasons one might put forward in answering this question. First, and almost painfully obvious, is that the despoiling of the natural environment—the pollution, overcrowding, and misuse of our resources—is more widespread and visible to the general public today than in the past. Beaches and lakes in many areas are grossly polluted; campgrounds are overcrowded—partially a result of population pressure. Exhaust fumes in the city, coupled with thermal inversions and smog, as well as pollution accidents (such as the offshore oil slicks in California or the *Torrey Canyon* disaster in England), are likely to become more prevalent. In addition, due to our knowledge of the effects of cigarette smoking on the body, we are more conscious of the health hazards of bad, polluted air (*see* René Dubos, *Man Adapting,* 1965).

A second factor—more important because possibly it is little recognized—is the paradoxical fact that the traditional goals of the "old conservation movement" have been too narrow and are now questioned because they have been surprisingly successful. Nothing can kill a social movement more quickly than success. A successful movement easily loses its élan, its energy, and it is apt to turn to the problem of keeping the bureaucracy going, administering the plans already made. Something of this order has happened with the United States Forest Service, one of the great achievements of the old conservation, which is well described in Herbert Kaufman's book, *The Forest Ranger: A Study in Administrative Behavior,* 1960.

The organizational structure of the Forest Service, as described by Kaufman, is a kind of hybrid offspring of the Methodist Church, the Jesuits, and the U.S. Army. This has led to a very efficient organization, but it has also meant that creative, innovative minds concerned with conservation issues in the general society were apt to look elsewhere, outside the official bureaucracies, for a place to express their concerns. I believe that it is these people who have become, in many instances, the leaders of the new conservation.

The old conservation movement was dominated by two traditions. One, symbolized by Theodore Roosevelt and Gifford Pinchot, was concerned mainly with the preservation of America's forestland (and obvious cash crop). Pinchot introduced scientific forestation to America, and the legacy of this effort is the U.S. Forest Service (the Department of Agriculture). The second, symbolized by John Muir, was concerned with scenic preservation and the establishment of parks. Its great legacy to the American people is the National Park Service (Department of the Interior).

Both aspects of the old conservation movement have been highly successful and, in principle, have won most of their major battles. The Forest Service introduced scientific forestry across the United States. A few decades ago it was fashionable to talk about the coming "timber famine," but it is now clear that this is no longer a central issue. (There is some evidence that at one time the Forest Service published low timber estimates in order to pressure Congress into appropriation of more funds for research.) Today the Forest Service holds millions upon millions of acres of land in public trust, much of it to be developed for recreational purposes. The same success story is characteristic of the National Park Service. We have perhaps one of the best-run and most adequate national park systems in the world. There are, of course, continual problems such as population pressure on campgrounds (Yosemite is a classic case) and a few battles to be fought over specific park sites (as the recently proposed North Cascade Park). Yet the truth remains that given the traditional goals of these federal agencies—the preservation of forests and the protection of scenic wonders—these two agencies have fulfilled their mission well, and much of their time, budget, and staff energy now goes to the tasks of administration.

As we have said, for a social movement to be successful it needs articulate spokesmen. Furthermore, in the case of environmental contamination issues, such leaders must be recruited from those people who are objective, competent, and have some grasp of the scientific problems involved; they cannot just be emotional preachers of a "hot gospel." In the past, the responsible leaders of the old conservation were generally produced by America's schools of forestry (a contribution to American education usually ignored). Since the American university has been, in many cases, highly innovative and responsive to public need (in particular, the land grant colleges), the question arises as to where the leaders of the new conservation will come from in the future, what kind of experts we will need, and where they may be most adequately trained.

In the journal *Daedalus,* 1967 (The American Academy of Arts and Sciences), several writers for a symposium of "America's Changing Environment" suggested that we need a new kind of trained expert (called by some "environmental manager"). No one, however, said very much about how they should be trained and under whose auspices in the university. They left the situation open to the typical (and often tragically wasteful) response of American higher education—creating a new series of institutes or special departments.

The situation is somewhat similar to what it must have been in city planning a few years ago. Everyone recognized the need for city planners, but no one wanted to integrate the existing strengths and expertise of the already established disciplines within the university—law, sociology, architecture, etc. As a result the university, like some uncontrolled amoeba, spawned a variety of institutes of city planning; along with this came an increase in internecine warfare between departments and often an unnecessary duplication of courses (e.g., urban sociology), a bewildering number of certificates of certification (M.E.D., Master of Environmental Design; M.C.P., Master of City Planning; and M.U.S., Master of Urban Studies), and in some cases, a lowering of academic standards.

The construction of new academic institutes and departments (better called fortresses) is dandy for inflating academic budgets, creating demands for new buildings, and establishing homes for wayward scholars and erstwhile administrators. But can we afford this typical, free-wheeling, academic kingdom building when confronted with the problems of environmental contamination? I think not. The problem is already very much here and now, so we do not have much lead time. Also, because of the Viet Nam war and scientific commitments already made (especially in the federal budget), we must capitalize on the already existing human resources and institutions.

Another factor is that the schools of forestry and conservation in American higher education are already innovating curriculums and programs to meet the crisis. Furthermore, by history and tradition they are uniquely qualified to provide the kind of expertise needed for today's environmental problems.

For example, consider the Yale School of Forestry. Strictly a graduate school, it was founded in 1900 when Yale was already 199 years old, and embarked on a pioneering venture to train men for a profession—forestry—that did not exist on this continent. The Yale School has flourished and has survived to become the oldest forestry school in America. When the school was founded by Gifford Pinchot and Henry Graves, they made certain the curriculum emphasized the biological sciences while also stressing problems of management and economics. Thus there has developed at Yale a tradition as well as an institutional structure whereby the social sciences and biology are expected to work closely together. The Yale model became the prototype for most schools of forestry. Furthermore, because of the nature of forestry as a profession, the men working in it were forced to combine scientific knowledge with an ability to work with federal and state governments, and business. Traditionally, to be successful, one needed to be a good botanist, a zoologist, and a geologist, as well as a good administrator and politician. These are exactly the combination of skills needed to meet our present environmental crisis, which is itself caused by a combination of biological factors, sociological conditions, and administrative entanglements.

A fourth factor encouraging the emergence of the new conservation is the propaganda success of the old conservation. Public relations are the *bête noire* of the American intellectual. But public relations techniques may be used for worthwhile purposes. It is a fact that the conservationist movement has been successful; today, to be "against conservation" is almost like being against motherhood.

I have studied the content of public school children's magazines such as *The Weekly Reader.* They are filled with stories concerning our space program (propaganda straight from NASA) but they also give a big pitch (and have for years) to conservation. There are many simple stories on animals, clean water, cleaning up the campground, and other traditional Boy Scout themes. If this all sounds a little sentimental, let me assure you that I do not look at it in this way. When I think of the child who may grow up to build a rendering plant on the river that runs near my property, or who, when of age, will drill oil off the coast of California and bring in big wages, I hope (perhaps fortuitously) that somewhere along the line he has had some exposure to conservation thinking. In any case, the conservationists have been fairly successful in winning over the NEA groups and popularizing their ideas in the public schools.

Another indication of the success of the message of the old conservation is represented in the field of mass communication. Perhaps the greatest conservation propaganda success of all time is to be found in "Lassie" and "Gentle Ben"—shown every Sunday night on CBS-TV. (Unfortunately, we hear that "Gentle Ben" is to be discontinued after this season; let's hope that it will be replaced with a show as pertinent to conservation interests.) One might argue that in the "Lassie" show the Forest Service has involved itself in the greatest publicity coup since spinach discovered Popeye.

There are other signs that the traditional conservation movement has been relatively successful in making the general public aware of the need for conservation. The recognition of the general problems of pollution and environmental degradation are, however, speeding up this communication process and spreading an awareness of public issues to new segments of the population—as in the case of black power. Thus the new conservation builds on the achievements of the past but extends itself, as a nascent social movement, into the realms of democratic participation and the new science of ecology. It continues the valuable tradition of conserving the world of nature; but carries the process into another social and political realm. It deals with a new metaphysics, a new view of the world that, based on the science of ecology, forces the conservation movement into a much more critical stance vis-a-vis the culture (industry, social change, etc.) than was true of the past.

Briefly, the ecological concept stresses that the human environment is a complex web of interacting variables, and many of these variables are the facts of the natural world (biological and physical). Thus, a change in one factor (for instance, bad air) has ramifications throughout the human environment—both biological and cultural effects. Furthermore, if one understands the connections, or pathways, of the system (energy changes), one may be able to predict over-all results of certain social actions (such as putting more DDT into the environment).

Some of the recent activities of the new conservation groups have come from the leadership of scientists and other specialists. For example, the Environmental Defense Fund of New York has (in Michigan and Wisconsin) been sitting in federal and state courts to establish the right to a clean environment. They have focused on the use of hard pesticides in the environment (for instance suing in Michigan courts to discontinue spraying the Dutch elm with DDT). Other groups, such as the Committee of Scientists for a Quality Environment in Connecticut, which was established to protect the marshlands of the Connecticut shore, have been formed by leading ecologists and although founded by professionals, they have received considerable public suport.

Such efforts seem to engender this support. In Michigan there was widespread public reaction, almost all favorable, to the efforts of the Environmental Defense Fund. Out of this has come, in several communities in the state, the spontaneous organization of local environmental action groups.

In Grand Rapids, Michigan (a city of 200,000), an environmental action program was started, linking ecological scientists and the public.

With the support of local business, labor unions, and newspapers, an Environmental Information Center was set up for western Michigan, pressure was exerted on the State House (the governor appointed a committee to study pesticide use in Michigan) and on local business to give attention to water and air pollution. A campaign also was launched to inform the public on the danger of accumulating hard pesticides in the environment. Attending some of the original meetings of this group, I was impressed by the enthusiasm and widespread concern shown by the Grand Rapids organization and the people involved. The truly striking thing, however, and a point corroborated by a survey of community leadership and general households, is that concern with the issue of the environment seems to cut across the usual sociological dichotomies. This was especially evident in the political realm where people, ranging from rather conservative to liberal orientations, could join forces and work together on local environmental issues. I was also impressed by the wide range of occupations, from working man to professional, that responds to this issue (as the sociologist would put it, the issue seems to cut across class lines).

Thus, after three years of studying citizen conservation groups in the Midwest as a participant observer—attending hundreds of meetings, listening to discussions, and conducting informal interviews—I am deeply impressed by several aspects of the new conservation. One is, of course, its strong moral sense. Sociologically speaking, if social thinkers from Emile Durkheim to Talcott Parsons are correct that a public acceptance of underlying social values is essential for maintaining social order, then the emphasis on moral indignation characteristic of many participants in the new conservation is an important fact. Secondly, if a politics of consensus is somehow important—in order to keep the social structure functioning—then it is of some interest that on this issue it is possible for extreme conservatives and liberals to work together. The politics of confrontation and conflict may not be the last word. There may be, and I consider this a hopeful sign, many areas of agreement in American political life that transcend the current issues of confrontation, and give some hope for a sense of common purpose and a more livable country.

2

"The Economics of Wilderness"

by Garrett Hardin

Garrett Hardin is Professor of Biology at the University of California at Santa Barbara. Concerned with what he terms the "humanistic bearings of biology," Dr. Hardin began his career in the field of experimental microbiology, moved toward genetics and evolution, and is now involved with questions relating to the cultural, ethical, and emotional values of scientific thought. He is the author of Population, Evolution and Birth Control *(1969),* Nature and Man's Fate *(1959), and* Birth Control *(1970).*

To some it may seem anathema to mention wilderness and economics in the same breath. Certainly, in the past, some of the most dangerous enemies of wilderness have been men who spoke the economic lingo. Despite this historic tar I think the brush of economics is a proper one for painting a picture of wilderness as a problem in human choice.

Economics may be defined as the study of choice necessitated by scarcity. There is something odd, and even improper, in speaking of the "economics of abundance" as Stuart Chase once did. With true abundance all economics ceases, except for the ultimately inescapable economics of time. Of the economics of time there is no general theory, and perhaps cannot be. But for the *things* of the world there is an economics, something that can be said.

Although there really is no such thing as an economics of abundance, the belief that there is, is one of the suppurating myths of our time. This belief had its origin partly in a genuine economic phenom-

This article is based on a talk given by Garrett Hardin at the Sierra Club's Wilderness Conference, San Francisco, March 15, 1969.

enon, "the economy of scale." For complex artifacts in general the unit cost goes down as the scale of manufacture increases. In general, the more complex the artifact, the more striking the economy of scale: the cost per unit to build a million automobiles per year is far, far less than the cost per unit when only one is manufactured. Because artifacts are so pervasive in modern life, most of us unconsciously assume "the bigger the better," and "the more the cheaper." It takes a positive effort of imagination to realize that there are things the supply of which cannot be multiplied indefinitely. Natural resources in general, and wilderness in particular, fall in this group.

This is obvious enough to Sierra Club members. It should be obvious to everyone, but it is not. Not long ago, for example, discussing some proposed improvements in a national park, the *Toronto Financial Post* said: "During 1968 and early 1969, campsites will be expanded and roads paved to enable the visitor to enjoy the wilderness atmosphere that was nearly inaccessible only a few years ago." This is an astonishing sentence, but I will bet that one would have to argue with the writer of it for quite a while before he could be made to see the paradox involved in speaking of building a road into the wilderness.

Wilderness cannot be multiplied, and it can be subdivided only a little. It is not increasing; we have to struggle to keep it from decreasing as population increases. Were we to divide up the wilderness among even a small fraction of the total population, there would be no real wilderness available to anyone. So what should we do?

The first thing to do is to see where we stand, to make a list of possibilities without (initially) making any judgment of their desirability. On the first level of analysis there are just three possibilities.

1. The wilderness can be opened to everyone. The end result of this is completely predictable: absolute destruction. Only a nation with a small population, perhaps no greater than one percent of our present population, a nation that does not have at its disposal our present means of transportation could maintain a wilderness that was open to all.

2. We can close the wilderness to everyone. In a limited sense, this action would preserve the wilderness, but it would be a wilderness like Bishop Berkeley's "tree in the quad" when no one is there: does wilderness really exist if no one experiences it? Such an action would save wilderness for the future, but it would do no one any good now.

3. We can allow only limited access to the wilderness. This is the only course of action that can be rationally defended. Only a small percentage of a large population can ever enjoy wilderness. By suitably defining our standards, and by studying the variables in the situation, we can (in principle) work out a theory for maximizing the enjoyment

of wilderness under a system of limited access. Whatever our theory, we shall have to wrestle with the problem of choice, the problem of determining what small number among a vast population of people shall have the opportunity to enjoy this scarce good, wilderness. It is this problem of choice that I wish to explore here.

What I have to say applies not only to wilderness in the sense in which that term is understood by all good outdoorsmen, but also to all other kinds of outdoor recreational areas—to national parks, to ski areas, and the like. All of these can be destroyed by localized overpopulation. They differ in their "carrying capacity." The carrying capacity of a Coney Island (for those who like it, and there are such people) is very high; the carrying capacity of wilderness, in the sense defined by Howard Zahniser, is very low. In the Wilderness Bill of 1964 Zahniser's felicitous definition stands for all to admire:

"A wilderness, in contrast with those areas where man and his own works dominate the landscape is hereby recognized as an area where the earth and its community of life are untrammeled by man, where man himself is a visitor who does not remain."

The carrying capacity of Coney Island is, I suppose, something like 100 people per acre; the carrying capacity of a wilderness is perhaps one person per square mile. But whatever the carrying capacity, as population inexorably increases, each type of recreational area sooner or later comes up against the problem of allocation of this scarce resource among the more than sufficient number of claimants to it. It is at this point that the problem of limited access must be faced.

How shall we limit access? How shall we choose from among the too-abundant petitioners those few who shall be allowed in? Let's run over the various possibilities.

First: By the marketplace. We can auction off the natural resource, letting those who are richest among the sufficiently motivated buy. In our part of the world and in our time most of us unhesitatingly label this method of allotment "unfair." Perhaps it is. But don't forget that many an area of natural beauty available to us today has survived unspoiled precisely because it was preserved in an estate of the wealthy in past times. This method of allotment has at least the virtue that it preserves natural treasures until a better, or perhaps we should merely say a more acceptable, method of distribution can be devised. The privilege of wealth has in the past carried many of the beauties of nature through the first, destructive eras of nascent democracy to the more mature, later stages that were capable of appreciating and preserving them.

Second: By queues. Wilderness could be made available on a first-come, first-served basis, up to the extent of the carrying capacity. People would simply line up each day in a long queue and a few would be allowed in. It would be a fatiguing and wasteful system, but while it would be "fair," it might not be stable.

Third: By lottery. This would be eminently "fair," and it would not be terribly fatiguing or wasteful. In earlier days, the decision of a lottery was regarded as the choice of God. We cannot recapture this consoling belief (now that "God is dead"), but we are still inclined to accept the results of a lottery. Lotteries serve well for the allocation of hunting rights in some of our states where big game abounds.

Fourth: By merit. Whether one regards this as "unfair" or "fair" depends on the complexion of one's political beliefs. Whether it is fair or not, I will argue that it is the best system of allocation. Anyone who argues for a merit system of determining rights immediately raises an *argumentum ad hominem.* He immediately raises the suspicion that he is about to define merit in such a way as to include himself in the meritorious group.

The suspicion is justified, and because it is justified it must be met. To carry conviction, he who proposes standards must show that his argument is not self-serving. What I hereby propose as a criterion for admission to the wilderness is great physical vigor. I explicitly call your attention to this significant fact: I myself cannot pass the test I propose. I had polio at the age of four, and got around moderately well for more than 40 years, but now I require crutches. Until today, I have not traded on my infirmity. But today I must, for it is an essential part of my argument.

I am not fit for the wilderness I praise. I cannot pass the test I propose. I cannot enter the area I would restrict. Therefore I claim that I speak with objectivity. The standard I propose is not an example of special pleading in my own interest. I can speak loudly where abler men would have to be hesitant.

To restrict the wilderness to physically vigorous people is inherently sensible. What is the experience of wilderness? Surely it has two major components. The first is the experience of being there, of (in Thoreau's words) being refreshed "by the sight of inexhaustible vigor, vast and titanic features," of seeing "that nature is so rife with life that myriads can afford to be sacrificed and suffered to prey on one another; that tender organizations can be so serenely squashed out of existence like pulp. . . ."

The experience of being there is part of the experience of wilderness, but only a part. If we were dropped down from a line by helicopter into the middle of this experience we would miss an important part of the total experience, namely the experience of getting there. The exquisite sight, sound, and smell of wilderness is many times more powerful if it is earned through physical achievement, if it comes at the end of a long and fatiguing trip for which vigorous good health is a necessity.

Practically speaking, this means that no one should be able to enter a wilderness by mechanical means. He should have to walk many miles on his own two feet, carrying all his provisions with him. In some cases, entrance might be on horse or mule back, or in a canoe, or by snowshoes; but there should be no automobiles, no campers, no motorcycles, no totegoats, no outboard motors, no airplanes. Just unmechanized man and nature—this is a necessary ingredient of the prescription for the wilderness experience.

That mechanical aids threaten wilderness is already recognized by managers of our wildernesses. Emergency roads, it is said, should be used sparingly. I submit that this cautious policy is not cautious enough. I submit that there should be *no* emergency roads, that the people who go into the wilderness should go in without radio transmitters, that they should know for certain that if an emergency arises they can get no help from the outside. If injured, they must either somehow struggle to the outside under their own power or (if lucky) catch the attention of another rare wanderer in the wilderness and get him to help. For people who are physically prepared for it, the wilderness is not terribly dangerous—but such danger as there is, is a precious part of the total experience. The knowledge that one is really on one's own is a powerful tonic. It would be cruelly sentimental to take this away from the wilderness adventurer.

There is not even a public interest in making the wilderness safe. Making great and spectacular efforts to save the life of an individual makes sense only when there is a shortage of people. I have not lately heard that there is a shortage of people.

There is, however, a public interest in making the wilderness as difficult and dangerous as it legitimately can be. There is, I think, a well-founded suspicion that our life has become, if anything, too safe for the best psychological health, particularly among the young. The ever greater extension of the boundaries of legal liability has produced a controlled and fenced-in environment in which it is almost impossible to hurt oneself—unless one tries. The behavior of the young clearly indicates that they really try. Drag races, road races, "rumbles," student sit-ins, marches, and tauntings of the police—all these activities look like the behavior of people looking for danger. I do not wish to deny that some of the activities may arise from other motivations also, e.g., idealistic political beliefs. I am only saying that it looks like deliberate seeking of danger is part of the motivation of our obstreperous young. I think it is an important part. I think we would do well to tear down some of the fences that now deprive people of the possibility of danger. A wilderness without rescue services would contribute to the stability of society.

There is a second way in which the interest of society is furthered by a rigorous wilderness. From time to time a president of the United States endeavors to improve the physical condition of the average citizen by resorting to a rhetorical bombardment. The verbal ammunition consists principally of the words "responsibility," "duty," and "patriotism." These rhetorical duds no longer move the young. The negative motivation of shame is, in general, not as effective as the positive motivation of prestige. A wilderness that can be entered only by a few of the most physically fit of the population will act as an incentive to myriads more to improve their physical condition. The motivation will be more effective if we have (as I think we should) a graded series of wilderness and park areas. Areas in which the carrying capacity is reckoned at one person per thousand acres should be the most difficult to enter; those with a capacity of one per hundred acres should be easier, those with one per ten, still easier, and so on. Yosemite Valley should, I suggest, be assigned a carrying capacity of about one per acre which might mean that it could be opened to anyone who could walk ten miles. At first, of course, the ten-mile walkers would be a very small class, but once the prestige factor took effect more and more people would be willing to walk such a distance. Then the standard should be made more rigorous.

I am sure other details of such a system would eventually have to be faced and worked out. It might be necessary to combine it with a lottery. Or some independent, easily administered test of physical fitness might be instituted. These are details, and in principle can be solved, so I will not spend time on them. But whatever the details, it is clear that many of our present national parks and national forests and other recreation areas should be forever closed to people on crutches, to small children, to fat people, to people with heart conditions, and to old people in the usual state of physical disrepair. On the basis of their lack of merit, such people (and remember, I am a member of this deprived group) should give up all claim of right to the wilderness experience.

The poet Goethe once said, "We must earn again for ourselves what we have inherited," recognizing that only those things that are earned can be precious. To be precious the heritage of wilderness must be open only to those who can earn it again for themselves. The rest, since they cannot gain the genuine treasure by their own efforts, must relinquish the shadow of it.

We need not be so righteous as to deny the excluded ones all experience of the out-of-doors. There is no reason in the world why we cannot expand our present practice of setting up small outdoor areas where we permit a high density of people to get a tiny whiff of nature. Camping cheek by jowl with thousands of others in an outdoor slum does not appeal to me personally—I have not visited Yosemite Valley in thirty years—but there are people who simply love this slummy togetherness, a fact that Sierra Clubbers sometimes forget to find hard to believe. By all means, let us create some al fresco slums for the people, but not in the likes of Yosemite Valley, which is too good for this purpose. But, there will be little loss if some of the less attractive forest areas are turned into outdoor slums to relieve the pressure on the really good areas. We must have lakes that fairly pullulate with water skiers in order that we may be able to set aside other lakes for quiet canoeing. We must have easily reached beaches that fairly writhe with oily bodies and vibrate to a steady cacophony of transistor radios, in order to keep up other beaches, difficult of access, on which we can forbid all noise makers.

The idea of wilderness is a difficult one, but it is precisely because it is difficult that clarifying it is valuable. In discovering how to justify a restricted good to a nation of 200 million people that is still growing, we find a formula that extends beyond wilderness to a whole spectrum of recreational activities in the national commons. The solution of the difficult case erects a framework into which other cases can be easily fitted.

3

"Prosperity and the National Parks"

by Jack Hope

In his article, "Prosperity and the National Parks," Jack Hope, a field editor for Audubon *magazine, explores the effects of population and prosperity on the natural areas within the National Park System. A former senior editor of* Natural History *magazine, Mr. Hope has written many articles on conservation and national parks for major publications, including a piece in the 1970 supplement of the* Book of Knowledge. *A graduate of Cornell University with an M.S. degree in economics also from that institution, he has co-authored a book of environmental fiction with James K. Page, Jr., to be published shortly.*

Driving into the valley, you can see the falls. As the rushing water leaves the gray granite cliff to leap into space, it is whipped into frothy whiteness; and continuing its wild fall, the stream of white widens quickly, then hits the rocks below. The impact shoots the flying droplets upward and outward, but they fall again and instantly gather to slide and pound in a furious rush down the mountainside, pausing here and there to eddy and curl among unmoving boulders. At the lower falls, the white water holds together, and is driven deep into the valley floor before it glides away from the base of the 2,400-foot cliff.

For many millenniums, the view of the cliff, the valley, and Yosemite Falls changed only slowly. And, until recently, the landscape remained much the same as it was at the time of the last glacial retreat. But during the last century, the minds and words of men have given the region the status of first a state park, then a national park; and within this time, the soft green valley at the foot of the waterfall has undergone more changes than at any time in the last ten thousand years.

Today the valley contains two asphalt roads, three hotels, hundreds of cabins, an outdoor amphitheater, two grocery stores, a hospital, seven souvenir stands, a laundry, three auto service stations—and not much grass. On any given weekend between June and September, row upon row of tents butt up against one another, suburban fashion, and upwards of 40,000 visitors swarm over the landscape. At several locations, parking lots sparkle as the sun plays on the skins of several thousand of Detroit's latest creations. Lines of automobile traffic clog the access roads, while hundreds of other motorists impatiently scout the park for a vacant parking space.

In the evening, a haze hangs over Yosemite; not the fragrant mist of a wilderness valley, but a heavy layer of campfire smoke and exhaust fumes. Many of the daytime visitors depart, and the sound of human voices lessens somewhat; but the clatter of pots and pans, the snarl of a motorcycle, and the cacophony of transistor radios rise to take its place. Along toward morning, after campers have gone to bed, the wind through the evergreens and the age-old roar of the waterfall again become audible.

When you first drive into the park, you wonder how the valley used to look. And if, by chance, you happen to know that the act of Congress which established the National Park Service in 1916 states that the environment of the parks is to be preserved "unimpaired for the enjoyment of future generations," it occurs to you how futile this legislative idealism has become in only fifty years. But no wonder, for the language of the act incorporates a basic contradiction—the incompatability between the concept of "unimpaired," and the swelling ranks of the "future generations," which have now arrived in full force. And while the Department of the Interior is fully aware of its schizoid mandate, no one, from Park Service Director George Hartzog to Secretary of the Interior Stewart L. Udall, has yet decided how to direct the conflicting policies it engenders.

Probably the language of the organic legislation didn't strike anyone as being contradictory in 1916, when the Park Service first opened its doors to the public. But the population of the United States in those days was only 100 million, and of that number, only 356,000 found their way into the national parks.

For a number of reasons, park visitation increased slowly during the System's first thirty years. Few people owned automobiles and fewer still had the time (or courage) to drive them to the western states, where many of the national parks are located. Most of the population lived in a rural setting, and even life in the big cities wasn't particularly stifling. Air pollution was in its infancy, and the traffic jam hadn't yet come into its own. These was abundant open space. Woodlands surrounding metropolitan areas hadn't been leveled for shoebox developments or sliced with highways, and except in urban areas, few industries had shoved their ways to the riverbanks and lakeshores as a means of solving their waste disposal problems. If you wanted a swim in a cool mountain brook, a hike through an evergreen forest, or just a few days of solitude, you could find them relatively nearby, if not in your own backyard. The need for park space did not seem urgent, and in 1940 only 17 million people visited the National Park System.

But coincident with the unprecedented increases in population, and the mobility and affuence of the postwar years, the flow of traffic into the parks underwent a sudden upturn, from 33 million in 1950 to 79 million in 1960 and finally to the current figure of about 145 million man-days spent in the national parks each year. And if park visitation continues its geometric rate of increase throughout the remainder of the century, then during every day of the four summer vacation months of the year 2000, there may be an average of something like ten people on each acre of the National Park System. So, if you're planning to spend some of your retirement years camping in the national parks—even if you plan to camp in what is today called the "back country"—perhaps you'd better make your reservations now.

When dealing with magnitudes like these, when facing up to the pressures created by a wealthy and rapidly expanding society, you suddenly realize that it becomes entirely legitimate to ask whether the National Park System can survive prosperity.

Today, a typical visitor to the national parks has at least a two-week vacation and a total of almost four months of "off" time every year. He, along with 80 million others, owns an automobile, which in a matter of a few days can carry him over a series of superhighways to Yellowstone, Shenandoah, or any other park within the continental United States. He's probably saved enough to treat himself, his wife, and their 1.6 children to a trip of 2,000 miles, which may include several of the national parks. The cost of motel rooms may present something of a barrier, but if he owns a camp trailer or tent, he can cut expenses by staying overnight in a national park—provided he can find a vacant campsite.

If he is on vacation, trying, as many park visitors do, to see as many of the national parks as possible in a minimal period of time, his stay in each will be a short one—somewhere in the neighborhood of 36 hours. Of course, he can't get to see much of the park that way, nor can he get much of a feeling for the outdoors, other than what can be absorbed through the windshield of an automobile. But that's probably not the reason he came to the park in the first place, and so, like other visitors, he'll most likely spend 95 per cent of his time on less than 5 per cent of the total parkland—on the roadways and in the heavily developed and highly convenient areas surrounding the "visitor centers."

If you belong to this group of "windshield visitors," as Director Hartzog dubs them, your conception of the much talked about "park experience" might be typified by the remark of the well-dressed lady in Yellowstone's Old Faithful Inn who announced, as she sipped her gin and tonic, that she and her husband were "anxious to do the Tetons" before returning home to the East. "Do" in this case didn't include anything as rigorous as overnight camping, but then, most of the larger parks have hotel or motel accommodations, so probably her way of using the parks continued undisturbed.

Small wonder that those interested in the function of the parks worry over the nature of the benefits derived from a visit to the National Park System, for observation and statistics indicate that the park experience has taken on many of the characteristics of a circus sideshow. A three-minute look at Old Faithful ("watch it steam and spout!") or a quick jaunt on a boardwalk above the Everglades alligators ("living monsters from another era!") and the hurrying visitor is off to see another "freak" of nature in a different park.

On the other hand, the tourists who never stray far from the visitor centers are indirectly performing a service for which many Park Service employees are grateful. For, by confining their activities to a small portion of the park, these visitors greatly simplify the task of safeguarding the true wilderness areas where relatively few people venture. But perhaps this situation engenders a false optimism, for as park visitation increases, even wilderness areas will be subjected to destructive pressures.

Although the park experience has not been officially defined, a notion of what it might ideally consist of can be derived from a speech given by Director Hartzog in which he expressed the opinion that the "single and abiding purpose" of the National Park System should be "to bring man and his environment into closer harmony." The Director described the parks as "places of highest inspiration," having "scientific, cultural, esthetic" and "educational" values. Yet, the lady "doing" the parks through the window of the cocktail lounge, the overnight camper seeking to escape the cost of a motel room, and the weekending teenager who comes looking for age-mates and excitement somehow don't fit into this mold. They use the parks for reasons unrelated or only vaguely related to the richness of the natural environment enclosed within their boundaries; and the activities they pursue are largely those that could be performed elsewhere, with little loss to the pursuit, and with great benefit to the task of the National Park Service.

Which brings up the question, if you happen to be concerned over the welfare of the large natural parks, of the proper use of a park and if somehow the "improper" uses couldn't be screened out, without going so far as to administer lie-detector tests to incoming visitors. But then you recall that the National Park Service is a public-serving, self-perpetuating agency that depends, as all agencies do, on congressional appropriations. And appropriations, by and large, are dependent upon numbers—numbers of visitors, buildings, highways—easily counted physical numbers. Remembering this, it's easy to imagine that each year when the Director brings his case before the Bureau of the Budget, he must be somewhat grateful that he has not guarded his gates too closely and that park visitation has risen over the preceding year. The National Park Service is already a low man on the totem pole of budgetary priorities. Would it enhance their competitive position against such pork-barreling giants as the Army Corps of Engineers to greet Congress with the message that: "Through the exclusion of visitors whose recreation could easily have been obtained outside the natural areas of the parks, and by our refusal to destroy any more parkland for construction of parking lots and access roads, we have reduced visits to the National Park System by 2 per cent"? Probably not.

Furthermore, the National Park Service is wedded to the philosophy that a visit to one of the country's areas of outstanding natural beauty is a right and not a privilege. Anyone who happens to have a $7 Golden Eagle pass can visit any park, any place, and can pursue almost any type of recreation; the Superintendent doesn't ask if you are going on a nature hike or whether you are going to play poker somewhere among the canvas and metal shelters of the public campground.

And so, while the Park Service may on the one hand be rightly concerned because the park experience may not foster an appreciation, or even an understanding, of the natural environment, their actions might at times lead you to believe otherwise. Their policy to date has been one of accommodation; accommodation of rapidly increasing numbers of visitors, of automobiles, camp trailers, and other gadgets that the modern outdoorsman demands for "roughing it."

The meaning of *roughing it,* quite naturally, varies from park to park, just as it differs among people. At Isle Royale, a roadless half-million-acre island in Lake Superior, visited by only 13,000 persons per year, the practice probably embodies pretty much of what George Hartzog would describe as the "challenges and . . . rewards" of the wilderness. Here, the challenge might be summed up as survival in an environment that has not been hand tailored to minimize the reality of the struggle for existence. The reward, in addition to the sounds, sights, and smells of wilderness, might amount to the development of a greater appreciation of the forces that have been shaping man for the last 100,000 years, and perhaps a heightened awareness of the shaping that he, in turn, has performed during the last few centuries. Then, too, there is the thought that we most appreciate those things for which we work hardest. And this may well be the case at Isle Royale, where the park experience is now provided along the guidelines mentioned by the Director.

At any rate, a considerable number of people feel that as many as possible of the natural areas within the system can be managed so as to achieve the "single and abiding purpose" of building an awareness of the natural environment—even in such heavily visited natural areas as Yosemite and Yellowstone. They suspect that the appropriate aim of the National Park Service should be to make the parks as *different* from the urban environment as possible, rather than to provide the outdoors with "hot and cold running," and are skeptical of the real value of the night club entertainment, the $30 hotel rooms, and the souvenir stands made available in the parks. And, looking forward to the day when population has doubled or trebled 1968 levels, they somehow feel that the largest item in the Park Service's budget should not be "Construction, and Parkway and Roadway Construction $49.6 million in fiscal '68), but "Management and Protection" ($40.7 million).

It may still be difficult to appreciate just how prosperity—that condition to which we ascribe magical qualities—may have any sort of negative consequences or how it may have brought about deterioration of the park experience. For it is, in fact, the economic affluence of American society that has provided the time, money, and mobility to make your visit to the parks possible. In essence, prosperity has been a great source of freedom, and more than likely, you are grateful for your two-week vacation, and for your high-speed automobile and the multilaned highways, which contribute to it.

But after a number of vacation visits to any of the more popular national parks, the realization is thrust upon you that the workings of the same set of social forces have simultaneously removed an equally large amount of freedom and that the evidence of deterioration is abundant. In Yosemite Valley you find, not a pleasant outdoor experience, but a transplantation of suburbia, with all the accompanying ills of smog, noise, crowding, and juvenile delinquency. At the Grand Tetons you are unable to find a free site in which to park your lovely new camp trailer. Throughout the million acres of Everglades National Park, you discover the flora and fauna dying as a result of nearby drainage canals built to accommodate agricultural and real estate development. And on a drive over the Blue Ridge Parkway, connecting Shenandoah and Great Smokies national parks, you are caught in a two-hour traffic jam that is no more enjoyable than one you could have experienced back home on the Long Island Expressway.

Delving into these few manifestations of deterioration leads to the unfortunate conclusion that most of the problems of the National Park Service have their origins outside the boundaries of the park system. And although the problems may be alleviated (or compounded) by actions of the Service, the fact remains that they are equipped to deal only with symptoms and not with basic causes.

If this were not the case, if the Park Service were ordered by Congress to deal directly with the causes of their current headaches, then probably they could begin by dispensing birth control pills at the main gates of each of the national parks. But short of that, and lacking a popular appreciation of the consequences of overpopulation, the National Park Service must direct its attention to the other end of the problem and hope that it can continue to acquire land in adequate amounts to keep pace with the demands of the spiraling number of park visitors.

For this purpose, Congress in 1964 established the Land and Water Conservation Fund, which, it was hoped, would provide the primary source of funds from which new parklands could be purchased. Money flowing into this fund is derived in part from fees paid by the public for the privilege of using the recreation lands administered by the federal government. These lands include not only national parks but reservoir projects administered by the TVA, Army Corps of Engineers, Bureau of Reclamation, and three other agencies. The only problem with the fund is that it is inadequate. A recent study conducted by the Bureau of Outdoor Recreation concluded that in the next ten years the fund will fall $2.7 billion short of the financial requirements for needed land. And other facts complicate the issue. At this writing, for instance, there is before Congress a bill that would relieve the Army Corps of Engineers of the responsibility of collecting user fees on the reservoir lands it administers.

Meantime, the price of land still available for inclusion within the National Park System continues to rise at something like 5 to 10 per cent per year. Since it takes Congress years to appropriate money for a park once a site has been recommended, land prices have a strange way of doubling and tripling in the time period prior to purchase. And, while Congress hesitates to purchase land that has undergone what President Johnson calls "artificial price spirals," the prices continue to rise and the region's suitability for national park status continues to decline as local interests hasten to make maximum use of the area's resources before the government finally takes possession.

With circumstances like these, it seems clear that potential parkland is now as inexpensive and pristine as it will ever be. But this fact has no apparent effect on the urgency with which Congress views the situation. The proposed Redwoods Park in California and Cascades Park in Washington, for instance, were suggested as national parks 50 years ago and 30 years ago, respectively, but have been ignored until recently when the aggressive campaigns of conservation groups brought the issues to the forefront.

At times, if you follow the disputes over establishment of new national parks, you wonder if perhaps the poor bargaining position of the National Park Service isn't a manifestation of something larger than price disputes. In the case of the Redwoods Park, for example, the Park Service in 1964 gave top priority to the Redwood Creek site after conducting a survey of possible park locations. Sometime later, however, after lumbermen had voiced opposition to the preservation of this area, Secretary Udall suddenly announced that he had decided instead to champion a park in the Mill Creek region, which had already been heavily cut over. In his explanation for the change of heart, Udall noted that he "wanted to pick a park, not a fight," which might make you wonder if the intention of the National Park Service to set aside the finest scenery for the benefit of the American people was being accorded a lower priority than the economic interests of West Coast lumber firms.

Udall's dislike of fighting apparently is not shared by either conservationists or lumber companies who now employ full-page newspaper ads to voice their opposing points of view in the bitter struggle over the redwoods. But while they battle to maintain a favorable public image, timber firms are still using their saws on the redwoods, pending a congressional decision on the boundaries of the proposed park.

There is an important instance of the Secretary changing his mind in an opposite direction. In 1965, the U.S. Government Printing Office (whose expenses are borne by taxpayers) published thirty pages of glossy finished color photos and text advocating construction of two hydropower dams on the Colorado River. Marble Canyon and Bridge Canyon dams would be built by the Bureau of Reclamation as a money-making project, and would, as the book implied, provide the least expensive source of power for southwestern states. Among the points listed in favor of the dams was that the man-made lakes thus created would somehow "bring man a little closer to God."

For some reason, the book skimmed over the violent public opposition to inundation of 150 miles of the 10-million-year-old Colorado gorge and to impingement by one of the lakes on both Grand Canyon

National Park and Grand Canyon National Monument. Alternative means of providing the power—nuclear or coal-driven plants—were regarded as "confused concepts." Written contributions to this unsubtle document were made by the Commissioner of the Bureau of Reclamation and by Secretary Udall.

After a prolonged and colorful battle, one seasoned group of conservationists proved, to the apparent satisfaction of the administration, that one of the confused concepts—coal-powered plants—could supply the needed power at less cost than could the hydro plants. Early in 1967, Secretary Udall agreed that perhaps the dams were not really necessary after all. The proposal was laid aside, to the relief of those who suspected right along that the man-made lakes might not, in fact, bring the nation any closer to God. (In all fairness, and using the common conception of God's geographical location, it should be admitted that a few waterskiers stood a chance of being brought closer to Him by about ninety feet—the distance by which the dam advocates wished to raise the level of the Colorado River.) In any event, if you are inclined to worry over the money we wasted in printing the hydropower propaganda, your concern is probably unfounded—the booklets will most likely be used when the proposal comes up again.

In this instance, a spectacular display of some of the world's oldest geology, including portions of two units of the National Park System, escaped destruction by a hairbreadth, largely on the basis of that most potent of criteria—maney. It makes you wonder if all irreplaceable resources will eventually be subjected to the test of the marketplace, as if they were refrigerators or automobiles. You may even question the claim to sophistication of a society that embraces materialism as the hallmark of the good life.

And perhaps you ponder the fate of the whole of the National Park System. How "inviolate" will be the boundaries of Rocky Mountain National Park if a rich oil-shale deposit is revealed to a nation of 400 million persons owning 300 million automobiles? To 600 million owning 450 million automobiles? Who will question commercial intrusion, providing a "need" for the oil can be established?

"Need" is a concept often employed in disputes over the national parks. You are told, for instance, that the 35,000 annual visitors to Mount McKinley in Alaska need a wider, straighter road to provide easy access to the park. (Director Hartzog has even advocated helicoptors and "motor nature trails" as a means of access to the wilder areas of some parks.) The present road into the park, with minor modifications, is more than adequate to handle visitors, but is being widened regardless.

In ten years, when visitation at McKinley has expanded, largely as a *result* of the road, the statistical increase will be quoted to justify construction of new parking lots and mass camping facilities needed by the new visitors. And in twenty years, increased visitor density at the original visitor center and resulting damage to the landscape will dictate the need for a new road into another part of the park so that the camping pressure can be dispersed. And on and on, the "need process" repeats, not only within the parks, but, to a much greater extent, in areas adjoining their boundaries.

Statistical "proof of need," for new highways, dams, power lines, mineral leases, grazing lands, parking lots, residential space, and farm land, is easy to come by. Each year the chambers of legislative committees are inundated with this "evidence" in attempts to obtain authorization for a variety of park-related projects that involve substitution of man-made, mass-produced devices for portions of irreplaceable natural environment. But each of these projects is underlain by a value system, a set of assumptions seldom explored with the same painstaking care with which congressmen scrutinize statistics. It might be worth restating the "need" involved in some of the more troublesome proposals that have faced the National Park System, in an attempt to provide a rough framework for consideration of such projects as will arise with increasing frequency in the future, and to strip away some of the superficiality that mere statistics sometimes imply. If you were to undertake this project for one or two representative situations, your prospectus might read as follows:

We feel the need for a drainage canal close to the boundary of Everglades Park, under the assumptions that (1) the residences to be located on the land thereby reclaimed could not be situated elsewhere (as they can), (2) the speculative profits derived by local real estate firms outweigh the loss to the country as a whole of the consequent destruction of portions of the park, (3) the lives of any creatures thus destroyed are singularly unimportant, and (4) the ecology of regions surrounding national parks can be disregarded insofar as protection of the park is concerned; or, in a more general sense: We propose that cutting of timber be permitted in Olympic National Park, based upon the beliefs (1) that the proliferation of material objects thus created is of greater value to the American civilization than are any of the esthetic and nonmaterial contributions of the natural environment, (2) that the mere fact that the country is endowed with a rich environment dictates that resources shall be exploited to the fullest, and (3) that derived economic gains shall be transformed into additional population growth, which in turn will justify further exploitation.

Somehow it sounds different this way.

4

"Landscape Esthetics"

by Luna B. Leopold

Luna B. Leopold, a Senior Research Hydrologist with the United States Geological Survey, has worked principally in the western states—including Hawaii and Alaska. He has degrees in physics and meteorology, civil engineering, and geology. Although, in recent years, Dr. Leopold's primary concern has been geomorphology—especially river mechanics and sediment movement—his research has included studies in climatology, soils and hydrology. The author of more than 100 scientific papers, he is co-author of The Flood Control Controversy *and* Fluvial Processes in Geomorphology.

There are an increasing number of bills before Congress that in one way or another affect the landscape or the environment. Each of these requires seemingly endless numbers of congressional hearings, which are recorded upon endless reams of paper.

And if, for some reason, you happen to read the voluminous testimony surrounding one of these environment-affecting proposals, you will generally find a marked contrast between the volume and kind of information presented by those who are pressing for technical development—building a dam, constructing a highway, installing a nuclear power plant—and the testimony of those who either oppose the development or wish to alter it in some way. The developer usually employs numerical arguments, which tend to show that there is an economic benefit to be obtained by constructing something—whatever that something may be. The argument is usually expressed in terms of a "cost-benefit ratio." It is typically argued, for instance, that the construction cost of a given project will be repaid over a period of time and will yield a profit or a benefit in excess of the development costs by a ratio of, let us say, 1.2 to 1. The argument is further supported with great numbers of charts, graphs, tables, and additional figures.

In marked contrast, those who favor protection of the environment against development are fewer in number, their statements are based on emotion or personal feelings, and they usually lack numerical information, quantitative data, and detailed computations. Perhaps this is the reason why this latter group seems to be continually fighting rearguard actions—losing battle after battle.

The time has come when the argument of the environmentalist might best be presented by (1) separating facts from emotions in relation to the environment, and (2) by providing him with a means of quantifying his arguments: using numbers to talk about the landscape. While to some of us this may be a little like using a computer to describe Shakespeare, it seems that society still has the right to have all aspects of any proposed development presented in a way that is as objective as possible.

One strategy used by environmentalists or conservationists to combat this paucity of statistical barter involves an attempt to describe society's interest in landscape integrity in monetary terms, which make a region's esthetic attributes appear to be similar to the kinds of benefits ascribed to the planned development. From this has arisen the unfortunate and unsound procedure of evaluating recreation in terms of what is called the "visitor day." This argument is based on the idea that the average visitor to a particular place spends one dollar a day (or some other amount) there, which he would not have spent had he not visited the spot, and that the enjoyment derived therefrom is in direct proportion to the amount of time or money spent in the given area. To me, this procedure misses the whole point of recreational activity since we know, by experience, that recreational enjoyment is by no means dependent upon either of these factors. Rather, it appears more sound to develop a way of directly describing the quality of the recreational experience. A first step toward this goal is to objectively describe the landscape itself, which the recreationist visits for enjoyment without regard to expenditure of money or time.

Toward this end, in August, 1968, I began a study designed to produce a method that would quantify the esthetic features of the environment so that the resultant data could be used in many planning and decision-making contexts. Such data could be especially useful when choices must be made among alternative courses of action. They would tend to provide a more prominent consideration of nonmonetary values to society.

The event that gave rise to this report was the application made several years ago to the Federal Power Commission for license to construct a hydropower dam in the Hell's Canyon area of the Snake River in Idaho. Three dams are already operative in the upper parts of Hell's Canyon, but there is still about a hundred miles of undammed river that retains a special character because it is not easily accessible. The river in this section flows through a deep, narrow gorge whose rocks and vegetation give it a particular grandeur. Proposals for damming of the Snake have been vigorously opposed by the country's leading conservation organizations. In the ensuing controversy over this issue, the following alternatives have been proposed: to dam another, nearby river instead of the Snake, to change the size and location of the proposed dam, to abandon the dam proposals altogether, and so on—the usual gamut of moves and countermoves.

The problem posed, then, was to determine objectively if (and in what sense) Hell's Canyon is indeed esthetically unique. The answer to this question would hopefully guide the decision as to whether the river should be dammed in this area. The study thus undertook to evaluate the factors influencing the esthetic appeal of Hell's Canyon. Toward this end, a comparison was made between Hell's Canyon and other river valleys in central Idaho. Then another comparison was made between Hell's Canyon and several well-known valleys in presently established national parks.

Three types of factors appear relevant to landscape esthetics. These groupings of factors and their "subfactors" are listed in table 1. The first group involves the physical features of an area—the presence of mountains and valleys, width of valleys, height and type of mountains. The second group includes those features that have to do with the region's biology, especially—in the case of river valleys such as Hell's Canyon—the vegetation near the stream and on the mountainsides, and the biology within the water itself. A third class encompasses what I have called "human interest factors." These are often more intangible than either the physical or the biological ones, but they are nevertheless influential in determining how the landscape impresses us. For example, if one is at the point on the Delaware River where George Washington is supposed to have thrown the silver dollar, that historical incident, however apocryphal, gives that place a distinct meaning. The phrase "Lincoln slept here" is of a similar sort—there is a human interest associated with the sites where certain phenomena exist or where unusual events have occurred.

Another set of circumstances, also related to the human interest factors, is the presence or absence of vistas or scenic outlooks. The many travelers who pull off to the side of the road at a turnout marked "scenic viewpoint" or "scenic outlook," are an indication that the ability to look from some vantage point across great distances, often to mountains or into far valleys, gives the landscape some special character—whether at a mountaintop or in a river bottom.

The question of access also falls within the realm of human interest. In the listing of criteria I have broken accessibility down into two parts: access to the individual, especially the hiker, and separately, mass use, meaning availability to motorized transport.

Human interest is affected by the general level of urbanization; it can make a piece of landscape more interesting or the opposite. The view from the Berkeley hills, for example, across the bay to San Francisco is made attractive by the skyline of San Francisco itself. In this case, the presence of the city seen from afar seems to make the landscape more interesting. On the other hand, in certain kinds of mountain country the presence of a great many cottages along a road may tend to detract from its inherent character, and have the opposite effect.

Within the list of human-interest criteria, I have included the term "misfits." One often remarks on how certain kinds of architecture fit into a particular landscape; probably one of the things that makes the Swiss mountain landscape so appealing, especially to American visitors, is that the type of architecture and the handling of building location seem to fit especially well into the particular environment. On the other hand, if you were to put a flashing neon sign advertising hamburgers on one of the Swiss chalets, it would be a cultural shock, and I would call it a misfit. So, also, in American landscapes one might be enjoying a drive through a rural countryside and suddenly be confronted with a tremendous dump of car bodies or even an obnoxious roadside billboard. These I consider to be misfits: they are out of character with natural surroundings. One does not mind seeing a large neon sign in the city nearly so much as seeing it in the countryside where it seems out of place.

The other two classes of factors are less complicated than the human-interest group. The physical factors are the easiest to measure in the field. Such factors as river width, river depth, and certain other characteristics require only a recording device or an elementary observation of the river channel. Under biological factors the list includes water color, turbidity, amount of algae, and the kind and extent of water plants, which are often indicative of stream purity or pollution. Under the three categories, a total of 46 criteria were chosen to describe a landscape's esthetic character.

After the factors were chosen, twelve river valleys in central Idaho, including Hell's Canyon, were chosen for evaluation sites that would have some potential for power development. Locations with such potential were selected in order to restrict the sites under discussion to those having something in common with Hell's Canyon. Each site was physically evaluated by standing at the edge of the river, thus providing uniformity in the way the observer looked at the environment. One could just as well have chosen evaluation sites that were part way up the valley sides, but this would have had the disadvantage of putting the observer at varying distances from the river.

At each site the checklist of 46 items was filled out. It can be seen from table 1 that most of the physical factors could actually be evaluated with some common unit of measure. Others, however, had to be estimated in terms of categories—erosion of stream banks, for example. In all cases there were five evaluation categories specified in the checklist. During evaluation, each site was described by assigning to each factor a number from 1 to 5, according to its physical, biological, or human interest characteristics. Where physical measurements were involved, the five categories varied in their span in an unbroken progression. For example, the five categories of river width were: less than 3 feet, 3 to 10 feet, 10 to 30 feet, 30 to 100 feet, and more than 100 feet. In this way, the categorization of a given site with regard to one of the factors could always be fitted into the category quantities.

One of the purposes of the study was to eliminate personal subjectivity in landscape analysis. Accordingly, the "evaluation numbers" for each of the 46 factors in the checklist serve a descriptive function only; evaluation number 5, for example, is not to be interpreted as "superior" to evaluation number 1, or vice versa. If a given site has a river width of more than 100 feet, our analysis does not rank this area above one whose river width is, let's say, less than 3 feet, but merely assigns different evaluation numbers to each of these locations.

The results of such a comparative study depend in part on the sites chosen for comparison. This being the case, another set of comparisons was made between Hell's Canyon and a series of rivers in four national parks of the United States. In this way I could find out not only whether Hell's Canyon is very different or rather like other river sites in Idaho, but also how the region compared esthetically with some of the great beauty spots that the nation has already recognized by giving them national park status.

Any scheme for comparing landscapes must rest on some philosophical framework. The philosophy underlying the scheme I used is the following: *Landscape that is unique either in a positive or negative way is of more significance to society than one that is common.* A place of

TABLE 4-1

IN THIS STUDY, THE ESTHETIC QUALITIES OF SIXTEEN RIVER VALLEYS ARE DESCRIBED BY MEANS OF THE 46-FACTOR CHECKLIST AT RIGHT— A FIRST STEP IN DETERMINING THE RELATIVE UNIQUENESS OF EACH SITE.

Factor Number	*Descriptive Categories*	*Evaluation Numbers* 1	2	3	4	5
	PHYSICAL FACTORS					
1	River width (ft.) (at low flow)	<3	3-10	10-30	30-100	>100
2	Depth (ft.) (at low flow)	<.5	.5-1	1-2	2-5	>5
3	Velocity (ft. per sec.) (at low flow)	<.5	.5-1	1-2	3-5	>5
4	Stream depth (ft.)	<1	1-2	2-4	4-8	>8
5	Flow variability	Little variation		Normal	Ephemeral or large variation	
6	River pattern	Torrent	Pool & riffle	Without riffles	Meander	Braided
7	Valley height/width	≦1	2-5	5-10	11-14	≧15
8	Stream bed material	Clay or silt	Sand	Sand & Gravel	Gravel	Cobbles or larger
9	Bed slope (ft./ft.)	<.0005	.0005-.001	.001-.005	.005-.01	>.01
10	Drainage area (sq. mi.)	<1	1-10	10-100	100-1000	>1000
11	Stream order	≦2	3	4	5	≧6
12	Erosion of banks	Stable		Slumping		Eroding large-
13	Sediment deposition in bed	Stable				scale deposition
14	Width of valley flat (ft.)	<100	100-300	300-500	500-1000	>1000
	BIOLOGIC & WATER QUALITY FACTORS					
15	Water color	Clear colorless		Green tints		Brown
16	Turbidity (parts per million)	<25	25-150	150-1000	1000-5000	>5000
17	Floating material	None	Vegetation	Foamy	Oily	Variety
18	Water condition (general)	Poor		Good		Excellent
	Algae					

19	Amount	Absent				Infested
20	Type	Green	Blue-green	Diatom	Floating green	None
	Larger plants					
21	Amount	Absent				Infested
22	Kind	None	Unknown rooted	Elodea, duck weed	Water lily	Cattail
23	River fauna	None				Large variety
24	Pollution evidence	None				Evident
	Land flora					
25	Valley	Open	Open w. grass, trees	Brushy	Wooded	Trees and brush
26	Hillside	Open	Open w. grass, trees	Brushy	Wooded	Trees and brush
27	Diversity	Small				Great
28	Condition	Good				Overused
	HUMAN USE & INTEREST FACTORS					
	Trash & litter					
29	Metal (no. per 100 ft. of river)	<2	2-5	5-10	10-50	>50
30	Paper (no. per 100 ft. of river)	<2	2-5	5-10	10-50	>50
31	Other (no. per 100 ft. of river)	<2	2-5	5-10	10-50	>50
32	Material removable	Easily removed				Difficult removal
33	Artificial controls (dams, etc.)	Free and natural				Controlled
	Accessibility					
34	Individual	Wilderness				Urban or paved access
35	Mass use	Wilderness				Urban or paved access
36	Local scene	Diverse views and scenes				Closed or without diversity
37	Vistas	Vistas of far places				Closed or no vistas
38	View confinement	Open or no obstructions				Closed by hills, cliffs, or trees

TABLE 4-1 (Continued)

Factor Number	*Descriptive Categories*	*Evaluation Numbers* 1	2	3	4	5
39	Land use	Wilderness	Grazed	Lumbering	Forest, mixed recreation	Urbanized
40	Utilities	Scene unobstructed by power lines				Scene obstructed by utilities
41	Degree of change	Original				Materially altered
42	Recovery potential	Natural recovery				Natural recovery unlikely
43	Urbanization	No buildings				Many buildings
44	Special views	None				Unusual interest
45	Historic features	None				Many
46	Misfits	None				Many

KEY: $<$ less than, $>$ greater than, $\leqq$ less than or equal to, / divided by great scenic beauty is of importance because of its scenic qualities. On the other hand, one could imagine a unique site which is extraordinarily unattractive—a large, neglected, pestilential dump, for example. This also has significance for society, but in the opposite sense.

Having obtained the checklist data from 12 river valley sites in the Idaho region, the next step was to compare the sites factor by factor in order to determine the relative uniqueness of each factor at each site. Let us take the river width, for example. As mentioned, with regard to this factor each site would be placed in one of five categories. Most of the sites had river widths falling into category 4—somewhere between 30 and 100 feet. A small number had widths greater than 100 feet, and a few fell in category 3, having a width from 10 to 30 feet.

To discover the relative uniqueness of each site's river width, we then determine how many among the 12 sites had river widths falling within each of the 5 categories. If it happened, for example, that there was only one river more than 100 feet wide, no other sites would share the 5 categorization, and this would determine what I call the "uniqueness ratio" of this one river. The uniqueness ratio for the river in question is equivalent to the reciprocal of the number of sites sharing the category value. The number 1, or unity, divided by the number of sites sharing the value, in this case 1, gives a uniqueness ratio for that particular site of 1.0 If there were two sites that shared category 5 in river width, each would be assigned a uniqueness ratio for river width of unity divided by 2, or ½ (.50). If, for a particular factor, all 12 sites fell in the same category, each of the 12 would be assigned a uniqueness ratio of 1 divided by 12, or .08.

By this method, then, each site had a uniqueness ratio for each of the 46 factors that were measured in the field. Adding uniqueness ratios for all 46 factors for a given site yields a "total uniqueness ratio." In this way, the total uniqueness ratios for the 12 sites may be compared one with the other; the higher the ratio, the more unique the site. This is a way of measuring numerically the relative uniqueness of each of the sites chosen for comparison. The results of this uniqueness ratio technique, when applied to the 12 Idaho river valley sites, are given in table 4-2.

There is a technical difficulty involved with the simple addition of the uniqueness ratios for the 46 factors at a given site. By the process of addition, each of the 46 factors is given essentially equal weight in determining the total uniqueness ratio. On further consideration of the list of 46 factors, one may decide that some are for more important than others, and therefore, for certain purposes, selected groups of the factors can be used for other analyses. Nevertheless, the uniqueness ratios for the 46 factors present a general means for quickly comparing a group of sites.

TABLE 4-2

"TOTAL UNIQUENESS VALUE" IS AN OBJECTIVE MEASURE OF HOW DIFFERENT EACH SITE IS FROM OTHER SITES STUDIED, WITHOUT REGARD TO "POSITIVE" OR "NEGATIVE" ESTHETIC VALUES. RIVER POLLUTION AT SITE 7, FOR EXAMPLE, MAKES THIS AREA RELATIVELY UNUSUAL AND GIVES IT THE HIGHEST "UNIQUENESS RATIO."

Site No.	Location	Total Uniqueness Ratio
1	Wood River, 6 miles above Ketchum	11.07
2	Salmon River, 1/4 mile above Stanley	11.00
3	Middle Fork Salmon River at Dagger Falls	11.87
4	South Fork Salmon River, near Warm Lake	13.93
5	Hell's Canyon, below Hell's Canyon Dam	16.09
6	Weiser River at Evergreen Forest Camp on Highway 95	11.17
7	Little Salmon River, 6 miles north of New Meadows	23.10
8	Little Salmon River, 4 miles south of Pollock	13.78
9	Salmon River, 2 miles below Riggins	10.25
10	Salmon River, at Carey Falls, 20 miles above Riggins	14.31
11	French Creek, 1 mile above junction with Salmon River	11.95
12	North Fork Payette River, near Smiths Ferry	10.21

The uniqueness ratio technique is objective in that it does not distinguish whether a given site is uniquely esthetic or uniquely unesthetic. The valley of the Little Salmon River near New Meadows (Site 7), for instance, was indeed the least attractive, most uninteresting, and unspectacular of the 12 sites surveyed. But because this site was different from the others in being a stream that was sluggish, algae-infested, murky, and slow flowing, it rated a high uniqueness ratio—the highest of all the 12 sites. It was unusual in a negative sense.

As can be seen in the table, the second largest total uniqueness score is Site 5, the Snake River in Hell's Canyon. In interpreting its rank order in total uniqueness score, one can say that Hell's Canyon was different from all others in ways that made it scenically interesting and therefore unique in a positive sense. Surprisingly, then, the two highest uniqueness scores represented the two sites that might be called the two opposite ends of the scale of esthetic interest.

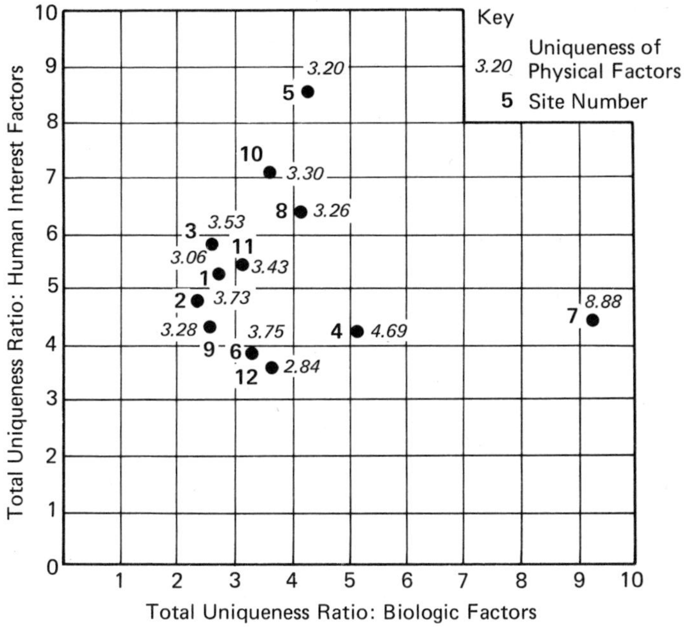

FIGURE 4-1: The graph provides a measure of the relative uniqueness of each of 12 Idaho river valley sites in terms of the three groups of esthetic factors listed in table 1—biologic uniqueness is measured on the horizontal axis, human interest uniqueness on the vertical axis, and the physical uniqueness ratio is listed alongside each site number.

A visual picture of the position of the various sites in the uniqueness scale can be obtained from the graph in figure 1 in which the ratios of human interest and biological uniqueness form, respectively, the ordinate (vertical, or *y*-axis) and abscissa (horizontal, or *x*-axis) with the physical uniqueness ratio written in numerical form next to each plotted point. It can be seen that Site 7 stands alone on the graph, primarily because of its uniqueness score on the biological scale—a result of its being the only polluted river among the sites surveyed. It can also be seen that Sites 5, 10, and 8 stand more or less alone because of low values in the biological factors and high values on the human interest scale. The graph, then, is an easy way of seeing that the total uniqueness score of Hell's Canyon, Site 5, and the polluted river, Site 7, are indeed at the opposite ends of a scale of landscape desirability.

Having demonstrated that the total uniqueness score does not involve personal preference or preference bias, one can now proceed to select combinations of factors from the checklist in order to perform additional types of analyses. In this study, checklist factors were chosen for their particular significance with regard to the impression that the site gives to the viewer. The *selection* of factors does involve personal judgment as to which ones appropriately describe the landscape characteristics; but the selected factors themselves remain independent of this judgment. Factor selection may therefore be thought of as a subdivision of the objective basic data.

In this step factors were selected from the 46-factor checklist for the purpose of evaluating each site in terms of two characteristics—valley character and river character. Valley character I conceived as a *combination* of the scale or grandeur of the landscape, the availability of distant vistas, and the degree of urbanization. Valley character is certainly influenced, in part, by the bigness of landscape features. The spectacular character of many of the grandest scenic views in the world comes in large part from the presence of high peaks in close proximity to the valley floor from which the viewer is seeing the landscape. Specifically, where the valley floor is narrow and the adjacent hills or mountains exceptionally high, the viewer has the concept of a large-scale landscape. Where, in contrast, the valley floor is very wide and the adjacent hills low, the impression is one of flatness—the opposite of grandeur. The Swiss Alps are so spectacular because the valleys are narrow and the mountains are nearby and extremely high. The same can be said of the Teton Valley, in the Jackson Hole area of Wyoming. One, therefore, can make a scale that shows the combination between two of the measured factors, the width of the valley floor and the height of the adjacent hills or mountains, in order to evaluate this grandeur aspect of each site, or what I have here called "landscape scale."

The method of obtaining a value for landscape scale for each of the 12 Idaho river values is dependent upon a diagram, as shown in figure 2. For each of the sites, the height of the mountains was plotted as ordinate and the width of the adjacent valley floor plotted as abscissa on the bottom of the diagram. On this graph, Sites 5, 10, and 11 fall in a zone of large-scale landscapes (high mountains, narrow valleys) and other sites fall near the other end of the scale, which I call a "subdued landscape." To simplify the combination of height of hills and width of valley to the single value, landscape scale, the position of each plotted point in the bottom graph in figure 2 is projected orthogonally onto a diagonal line (line A). (The projection of each point is indicated by those lines leading from the points at a 45° angle upward and to the right in the lower graph.) Line A now serves as the line on which the sites are ranked according to their respective landscape scales. Having

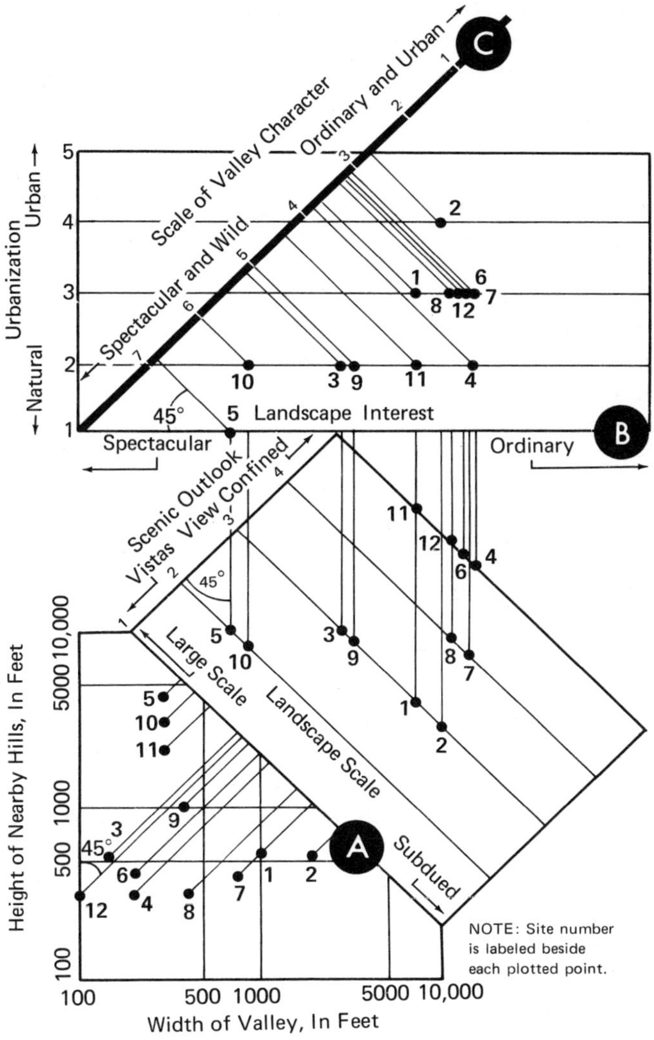

FIGURE 4-2: "Height of nearby hills" plus "width of valley" equals" landscape scale." This value plus "scenic outlook" equals "landscape interest," which, combined with "degree of urbanization" yields "scale of valley character," a measure of the viewer's esthetic impression of the landscape at each site.

a value of landscape scale for each site, this joint value can be used for the construction of a new graph, as has been done in the central portion of figure 2, using line A as a horizontal, or *x*-axis.

At this point, it was reasoned that the impression of a landscape on the viewer is partly determined by this factor just derived, landscape scale, and partly by the "degree of view confinement." Where distant vistas are available in large-scale landscapes, one has the impression of spectacular scenery. In contrast, where the view is confined by heavy cover or by adjacent hills, the result is esthetically ordinary from the scenic point of view. Again this is one of the reasons why the Alps give a strong impression of scenic grandeur; not only is one looking out from a narrow valley to the high mountains immediately adjacent, but one can see up and down the valleys for long distances, viewing ranges of mountains as background vistas. In contrast, when one is in the bottom of a narrow, tortuous gorge, with no distant outlooks, as in certain portions of the Black Canyon of the Gunnison River in Colorado, the impression of the view is distinctly less spectacular.

On the central graph of figure 2, then, scenic outlook (the presence or absence of distant vistas) serves as the *y*-axis and the landscape scale as the *x*-axis. And the position of each point on the graph is again projected upward at a 45° angle onto line B providing a scale which I have labeled "landscale interest." The position of each site on the landscape interest scale is thus a ranking of the individual sites using a combination of three factors: width of valley floor, height of adjacent mountains, and availability of vistas.

The scale of landscape interest can now be combined with another factor—degree of urbanization—which also affects the viewer's impression of the total landscape. The degree of urbanization is defined as the totality of buildings, houses, roads, utilities, and other earmarks of man-made change present in a given site. As before, using the scale of landscape interest as the *x*-axis, an ordinate, or *y*-axis, is constructed representing the degree of urbanization. This provides the final or upper graph in figure 2. The sites having a combination of spectacular scenery and minimal urbanization fall in the lower left-hand zone of the upper graph, whereas those having ordinary and urban conditions fall in the upper right. The combination of landscape interest and urbanization, each given equal weight, is represented by another 45° projection onto the diagonal in the upper graph (line C), yielding a scale of what I call valley character. We thus constructed a ranking scale called "scale of valley character" representing a combination of four of the factors from the checklist in table 1: width of valley, height of adjacent mountains, availability of vistas, and degree of urbanization.

When the position of each of the 12 sites is considered on this final scale of valley character, we see that Sites 5 and 10, Snake River at Hell's Canyon and the wild reach of Salmon River at Carey Falls, are highly unusual. They are characterized by narrow valley floors, high adjacent mountains, availability of distant vistas, and little or no urbanization. All of the other sites in figure 2, fall to the right of Sites 5 and 10.

After combining the four factors contributing to valley character, we can perform a similar analysis on the rivers themselves, obtaining a measure called "scale of river character." Experience indicates that the grandeur or majesty of a river is dependent upon a combination of its size and apparent speed. Rivers tumbling over a succession of falls tend to be more impressive or esthetically appealing than those that appear sluggish. The latter characteristic is not so much dependent upon true velocity, which is usually poorly estimated by the untrained eye, as it is on the *appearance* of speed judged mostly by waves and surface riffles caused by rapids or falls. We therefore wish to find a combination of factors that gives the viewer the impression of the river's grandeur itself. For this we use river width, river depth, and the presence or absence of rapids, riffles, and falls, chosen from the checklist factors.

Eventually we arrive at a ranking of the sites according to river character; those rivers that are wide, deep, and have rapids, falls, and riffles rank high on this scale. In this instance, Sites 5 (Hell's Canyon) and 10 (Salmon River at Carey Falls) again outrank the other locations. As the last step in our analysis, we use a graph to compare the 12 sites according to both valley character and river character (figure 3), and thereby obtain a final rank reflecting a total of 7 of the factors chosen from among the 46 on the initial checklist.

It can be seen that on this graph Hell's Canyon is indeed unique, for it falls in the farthest position in the upper right-hand part of the graph. Nearest to it again is Site 10, the wild reach of the Salmon River near Carey Falls. Nearly all of the other sites fall in the central cluster in the center of the graph. Site number 7 (Little Salmon near New Meadows), the least interesting of all those surveyed, falls at the opposite end of the scale, appearing in the lower left-hand portion of the figure.

Having evaluated the Idaho sites, a new comparison was made in which Hell's Canyon of the Snake was viewed along with rivers in four areas officially recognized as having great natural scenic beauty. These are the Merced River, in Yosemite National Park in California's Sierra Nevada; the Grand Canyon of the Colorado River; Wyoming's Yellowstone River near Yellowstone Falls in the national park; and the Snake River in the Teton National Park below Jenny Lake, again in Wyoming. How does Hell's Canyon compare with them?

The same kinds of data that were used in the construction of the graphs just described were tabulated for Hell's Canyon in combination with these four national park rivers. A similar set of graphs was derived, the results of which are shown in figure 4. In this comparison, it can be seen that the points representing Hell's Canyon and the Grand

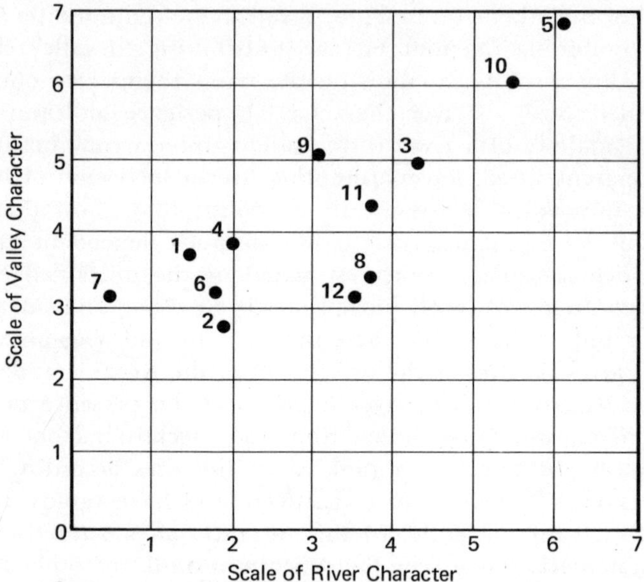

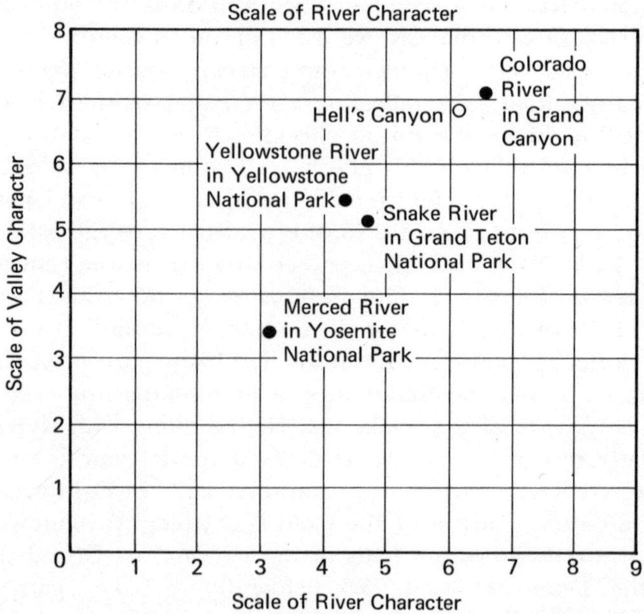

FIGURES 4-3 and 4-4: The viewer's esthetic impressions of Idaho river valleys and river valleys in national parks are evaluated in terms of "valley character" and "river character." Relative to other locations, numbered sites falling in the upper right-hand portion of the graphs tend to have large rapid-flowing rivers, large-scale landscapes, scenic vistas and little urbanization. On the basis of this assessment, Hell's Canyon ranks second only to the Grand Canyon of the Colorado.

Canyon fall nearly together in the upper right-hand portion of the graph. They are comparable in the combination of valley character and river character and stand in exceptional positions. By comparison, the Merced River in Yosemite is of a lower order of interest. Truly the Yosemite Valley as a whole is one of the great scenic beauty spots of the world, but the river within the valley is not of a special character. As far as the Snake River in Teton Park is concerned, its position in the center of the graphical comparison can be attributed in part to the higher degree of urbanization present there than in the Grand Canyon or Hell's Canyon, and to the comparatively large width of the valley floor, a characteristic that makes the Snake River at that point less of a grand spectacle than either Hell's Canyon or the Grand Canyon. Relatively great urbanization, along with the absence of scenic vistas, also diminishes the site of the Yellowstone River.

The result of the data collection and analysis indicates that it is possible to set up a list of factors that influence the esthetic nature of a given location. The factors can be considered all together, in this case by the computation of a total uniqueness ratio, or they can be selected and used in various combinations to express certain aspects of a landscape's characteristics. It is hoped that the study will indicate both the need for, and the possibilities of, objective description of landscape. Specifically, this analysis shows that the Hell's Canyon of the Snake River, the site proposed for a hydropower dam, has unique characteristics that give it an exceptional esthetic rating and place it in a category shared by a few other landscapes within the United States.

5

"Daniel Boone is Dead"

by David Lowenthal

David Lowenthal is a Research Associate and Secretary of the American Geographical Society. A visiting professor in the department of geography at the University of California, Berkeley, during 1969, he is a prolific writer and editor on conservation, geography, and history. Among his publications are "English Landscape Tastes" and "The American Scene" in the Geographical Review, *and* Man and Nature *(editor). Dr. Lowenthal, who has done extensive field work in the Caribbean Islands and Guianas as well as Britain and Western Europe, plans future studies in environmental perception.*

"Open space is like virginity," the general manager for state parks in New York City recently remarked. "Once lost, it can never be regained." But this is not entirely true of either. Non-virgins are needed to produce more virgins. And sanitary fill is a prime source of urban open space, if only because the settling of decayed matter precludes building for some time. Yesterday's dump is tomorrow's park, or, to paraphrase Wallace Stegner, a man's home is his castle, but his litter belongs to everybody.

Landscape is often seen as a commentary on the human condition. And as moral metaphors, landscapes are as American as Niagara Falls and redwoods, although not so durable. To New England Puritans the primeval forest was the locus of evil, terror, and witchcraft. But in the nineteenth century trees left the devil's domain for a higher place, and what with Arbor Day, Longfellow, Joyce Kilmer, and Smokey the Bear it is a wonder they are ever chopped down.

Moral purpose animates the design of both private and public space. A half-century ago, the United States Department of Agriculture advocated farmstead beautification on the ground that "a home and its surroundings must be attractive in order to be most uplifting to the family, visitors, and passers-by" (1920). Public parks were initially intended to restore the health, improve the character, and elevate the taste of the masses. Surrounded by blight and squalor, urban parks were places deliberately set apart—oases not only of greenery but of the behavior thought appropriate to greenery. "Foul air prompts to vice, and oxygen to virtue," was a truism of park planning in the 1890s.

Both open space and public opinion still reflect this point of view. Parks scarcely affect urban environment; they often seem pallid, woebegone substitutes for the country, rather than integral aspects of the city scene. The Secretary of Agriculture urges Americans to exchange the "rabbit warrens of the big cities for the fresh air, clean water, and space of the small towns." The latest *Yearbook of Agriculture* insists that "Man needs nature; . . . animals which are forced to lived under crowded conditions develop many of the antisocial traits that fill the crime pages of big city newspapers." The Department of the Interior endorses a Congressman's advice to get "beyond brick and mortar, away from the sound of cities . . . over the hills to God's own country . . . where health and happiness take root . . . where Nature quickens physical, mental and spiritual guidance."

Such pieties are regularly reiterated, but seldom documented. Some feel that evidence is unnecessary. Open-space preservation "should require no defense," asserts New York City's Open Space Action Committee (1965). Others take refuge in the ineffable: "If the rat and sparrow can learn to live for endless generations in the city, why cannot man?" asks Roger Revelle (1967). Scientists are unable to give us answers, he admits, "but the prophets and poets can."

Every conference on open space "ends up waving the green flag and we salute the Sequoia," a conservation educator complains. "Having saluted the Sequoia all my life, could we not now have a good reason?" When asked whether man had biological and psychological needs for natural experience, he confessed ignorance: "I don't even know whether the question is meaningful." Meaningful or not, the Department of Agriculture still entreats Americans "to create an outdoor environment that meets the high standards of living within most of our homes," because "the outdoors, too, is our home."

Is the outdoors in fact "our home," or is it viewed, like purity, as being a rare and precious commodity which to touch is to defile? In surveys I have made, people usually associate open space with arid lands in the Southwest, great expanses and distant horizons, sweeping plains and empty deserts. Cleanness and beauty and freedom predominate, but these places are devoid of human activity. In short, they are meant for reverence, not for recreation; play would pollute them. And so, by analogy, wilderness lovers feel that use desecrates the wild. They urge its preservation as an ideal, all the more precious because rarely, if ever, experienced. "We must ask even those who love the wilderness the most," says a prominent conservationist, "to touch it but seldom, and lightly."

The outdoor activities of wilderness lovers reflect these ideals. Let me sketch a few traits of the dedicated outdoorsman in order to point up the implications of his behavior for recreation and the landscape.

1. *The outdoorsman observes nature mainly to count and classify.* Birds are watched to be listed; mountains are climbed to be checked off on tally sheets; hikers vie with one another in terms of miles per day and pounds per pack; the Appalachian Trail is walked from end to end as an elite achievement. This is no parody of outdoor interests; such matters are the staple of conversation on many outings.

2. *The outdoorsman is there to improve himself.* Climbs, hikes, and canoe trips are learning experiences, and instruction is continuous. Sitting around the campfire in the evening, swapping tall tales? Not a bit of it. After the tents are up, supper over, and the gear stowed away, come lessons in knot tying, fiberglassing, life preserving, and equipment buying, or educational films on the finer points of whatever the group has just done or will do next.

3. *The outdoorsman is organized.* In the name of safety everything takes place in groups; teamwork is enforced on gentle trails just as on the steepest mountains. Being alone and getting lost were the height of wilderness experience for nineteenth-century visitors. Thus the painter Thomas Cole in the Catskills: "I was lost . . . I felt a wild and vivid pleasure . . . I shouted, sang, whistled, for the very horror of the thing." This is no longer fashionable. Recently I was exhorted to spare no effort in the design of a new edition of a New York area hiking guide "to prevent even one hiker or boy scout from getting lost or caught in the woods after dark."

The outdoorsman enjoys discipline whether he is dishing it out or taking it. When a novice is berated for some minor failure or infraction of the rules, he accepts it meekly; back home, such officious authority would be bitterly resented.

4. *The outdoorsman is a good citizen.* He is eager to be useful and cooperative, takes up the minimum space at campgrounds, is careful not to interfere with other campers, and leaves his site as tidy as a grave. Most other Americans still think of nature, in H. L. Mencken's phrase,

as "a place to throw beer cans on Sunday." But such atavism is a source of embarrassment to the serious outdoorsman.

5. *The outdoorsman is a masochist.* His prize recollections involve being attacked by mosquitoes or black flies, drenched by downpours or by falling into rivers, frozen by icy winds or snows, and semi-crippled by extreme exhaustion. Straining muscles and fevered temples are admired as products of "clean work! clean sweat!" but as one devotee confesses (or boasts), "few but the hiker will understand this reverence for exertion." A canoeist put it more baldly, describing portaging as "quite simple: just equal parts of masochism and brute force." And although paddling a heavily loaded canoe against the wind "is not exactly the same thing as being a galley slave, it does afford some insight into what life was like." Thoroughgoing outdoorsmen even make a fetish of hunger. We are enjoined to emulate Thoreau, who "boiled a handful of rock tripe . . . for more than an hour," and happily announced that "it produced a black puff . . . not positively disagreeable to the palate."

Discomforts are praiseworthy because they promote intimacy with nature; they also dissuade the tender and unconverted from putting in an appearance. Those who avoid discomfort are dismissed with derision or contempt. Neither age nor condition is an adequate excuse for refusing to suffer. When one outdoorsman noted for his ability to take a beating finally, in his sixties, bought a VW camper, his former cronies shook their heads over this mark of weakness, and the rumor went around that the poor man had slipped so far from the true path as to spend an occasional night in a motel.

Outdoorsmen who endure allow themselves only the most meager comforts and rewards. Novice climbers at the Appalachian Mountain Club winter training school are enticed to the summit by the promise of a lemon and a bite of "glorp"—a bar of pressed raisins, nuts, cereal, and chocolate. The shivering canoeist scathed by a river mishap is surreptitiously offered the smallest nip of whiskey to warm him.

I do not mean to traduce the experience or to impugn the motives of the outdoorsman. His ascetic traits notwithstanding, he must still be viewed as a recreation seeker, in the sense of Marion Clawson's definition of "recreation" as activity undertaken because one wants to do it. Nevertheless, his quest for purity and virtue in the outdoors is sharply at odds with the recreation demands and behavior of the great majority.

The general public views the outdoorsman as commendable, but seldom worth emulating. "The urge to 'get back to nature' is an important factor in the lives of increasing numbers of people," the United

States Forest Service claims, but its picture of what people do is sharply at variance with this statement. Most campers today are in a hurry; they have a lot to see and a schedule to meet. Equipment and facilities are increasingly luxurious. People with mobile trailers want electrical and water hook-ups rather than fireplaces and tables; they insist on hot and cold running water, showers, flush toilets, and laundry facilities. Indeed, the Forest Service concludes, "visitors seem to be increasingly 'soft.' They don't venture far from their cars. Life in camp, in terms of creature comfort, is not much different from that at home. . . . Even hunters are tending to use motels and restaurants as a base of operations instead of the traditional hunting camp." People may like a taste of the outdoors, but they usually do not want to live in it, however briefly.

Above all, most Americans are gregarious. Solitude and silent communion with the great outdoors are the last things the average camper seeks. "My notion of camping," writes Faith McNulty, "was that we'd . . . manfully make our little home in the wilderness, enjoying the slightly scary pleasures of solitude and independence." But finding no remote spots, they "camped where Americans are supposed to camp"— in state and national parks. "These camps are about as sylvan as Central Park. . . . The whole place was as busy and as merry as the zoo on Sunday. . . . When I mentioned the idea of solitary camping to any of our campmates, they looked puzzled or dismayed. 'You mean you want to camp where there's nobody *around?*' a woman in Zion Park, Utah, asked, in horror. 'Why, I'd be scared simply stiff! ' "

The traffic jams at popular national parks are deplored by nature lovers, but the ordinary tourist welcomes the presence of his fellows as a cheery reminder that he is not alone in the wilderness. At the Grand Canyon not one visitor in a hundred ventures below the canyon rim. Few stray from well-worn paths. In Yosemite Valley, Fourth of July campers are estimated at more than eight thousand per square mile: "The damp night air, heavy with a pall of eye-watering smoke, is cut by the blare of transistor radios, the clatter of pots and pans, the roar of a motorcycle, and the squeals of teenagers." Except for the trailers and tents "this might be any city after dark" ("Ah wilderness" 1966).

But this is essentially how many people like parks. Indeed, it is close to what was anticipated by the proponents of our first national park. When N. P. Langford admired Yellowstone Lake in 1870, he predicted it would soon "be adorned with villas and the ornaments of civilized life. . . . The march of civil improvement will reclaim this delightful solitude, and garnish it with all the attractions of cultivated taste and refinement."

Daniel Boone is reputed to have felt crowded when he saw the smoke from another cabin. Daniel Boone is dead. Most modern campers do not object to pitching their tents a few feet apart; a campground is said to be full only when you have to use the other fellow's tent pegs. Recreation specialists have conscious or unconscious standards of use intensity that most folk are willing, even happy, to exceed. After Labor Day, when most campers have left Yosemite, the remainder huddle together to maintain a comfortable feeling of density.

Most Americans enjoy nature not to "get away from it all" in the wilderness but to relax in familiar surroundings. The laundromat at Mesa Verde National Park made a lady from Kansas "feel right at home; they even have my favorite detergent." Modern camping is "the definitive means of getting close to nature . . . without coming to grips with it," according to Gilbert Millstein. "Nobody chews pemmican or gnaws on edible roots and only a handful of eccentrics . . . invite their souls on a bed of boughs." The campers at Groton State Forest, Vermont, all had portable refrigerators and were given stacks of firewood (from logs cut outside the forest). Millstein concluded that their pioneering drive was "a return to the soil—in the fashion . . . of Marie Antoinette pushing sheep around with a gilded crook in the Petit Trianon."

Not even proficient campers eagerly embrace all the rigors of the wilderness. Those interviewed at Glacier and Quetico-Superior national parks expressed pleasure in being able to cope with primitive surroundings, but their conception of "wilderness" was certainly not Aldo Leopold's. They all wanted to preserve primeval nature, but four out of five also asked for more campsites and amenities. They saw no inconsistency in wanting both. For many of them, a forest recreation survey concluded, "wilderness subsumes the existence of picnic tables, wells, toilets, washrooms, and the like."

The luxuries of camping are easy to lampoon. "Every time I unpack the folding chromium barbecue pit and the dehydrated soup mix, out there under the stars, I say to myself, 'Gil, boy, this is something you can't escape: it's in your blood' ". All-out primitivism holds little appeal for most Americans. They identify with Huck Finn, who carried as much equipment as possible: a frying pan, a coffeepot, tin cups, a knife, fishhooks, and a gun; his raft was not made of logs but of cut lumber. Hunters in the North Woods may like to identify with hardy eighteenth-century trappers, but bush airlines, light-weight equipment, packaged foods, and bottled gas make modern camping in the far north a luxury proposition. Comfort in camp, sporting operators assert, enhances the guests' enjoyment of the wilderness that surrounds them.

"As you grow older," observes a camper, "you learn to tolerate the comforts." Yet discomfort retains a certain cachet. One couple, tired of wet wood, burned pans, and canned rations, realized that they were not pioneers and were pleased about "being honest enough to admit this." They decided to doff their hair shirts and dined in restaurants each evening, but then had to endure the contemptuous disapproval of those who took camping more seriously. It might not be much of a challenge to survive on packaged food, but a camper ought at least to cook his own meals.

Today a modicum of effort suffices to derive virtue from nature. A redemptive outdoor experience demands no sacrifice of comfort, no great exertion of muscle; it is a voyage of the mind, a journey through forests that call less for the woodsman's lore than for the impresario's talents. The outdoor church at Table Rock Lake, in the Ozarks, features a biblical *son-et-lumière* entitled "The Shepherd of the Hills." On summer evenings in a California national forest campers sit around a simulated log fire listening to nature lore, watching movies, and singing oldtime ballads. After the performance, visitors join in the symbolic ritual of dousing the campfire, while rangers turn the valve that cuts off the gas. No embers smolder to cheer the campers' slumber. Tamed nature has played its passive role, and the modern Prospero returns home refreshed if not exactly reborn.

Every year, however, more and more Americans come back to the outdoors converted to the wilderness mystique. From hedonists and heedless despoilers they become serious-minded amateur ecologists, alive to the risks of pollution, angered by the Corps of Engineers, sworn enemies of the public utilities and lumber companies. And herein lurk new perils to the landscape. An elite minority today guards treasured wilderness against intensive recreational use by the unappreciative majority. But *enlightened* masses will enormously multiply the pressure on recreational land. No longer content to crowd together in Yosemite and Yellowstone, they will fan out to all the wilder reaches of the country. Under such an onslaught, the wilderness is apt to disappear, to be supplanted by fenced preserves and artificial "nature" areas for esoteric outdoor studies.

Intensive use by dedicated outdoorsmen already threatens some areas. In countrysides that easily accommodate individual ramblers, fishermen, and hunters, hiking clubs may be anathematized; organized recreation is apt to endanger or foreclose covenants with landowners allowing rights-of-way across private property. A dozen well-booted hikers as a body are more likely to disturb the vegetation and compact the soil of a trail than as many sauntering individual wayfarers. The organized group is also a visual, often an aural, outrage to the resident population and lends to the landscape an unwelcome impression of public ownership. The outdoorsman's environmental needs are thus both special and exclusive; coexistence with residential land use is almost out of the question. Both his tastes and his impact on the land push him away from lived-in countryside to wilderness reserves.

It is well to recall that the virtues of the wild were formerly seen in quite another light. Americans used to enjoy the wilderness, if at all, mainly to imagine how much better it would look when the forests were replaced by fruitful fields and teeming towns. Not until the nineteenth century did James Fenimore Cooper and the Hudson River school of painters provide a true wilderness esthetic. Even so, it was the association of wilderness with ruins that endeared it to many Americans. Dead trees were admired as analogues of European architectural relics, and ruined arches were placed up and down the Hudson to enhance the effect of nature.

The Catskills and Adirondacks were later esthetically outclassed by western landscapes more stupendous and more evocative of human associations. The stratified sedimentary and volcanic rock formations of the West moved visitors profoundly because they reminded them of ruins of great cities, castles, temples, and other monuments of ancient civilization. "One could almost imagine," wrote the geologist Frederick V. Hayden after his Yellowstone expedition of 1871, "that the idea of the Gothic style of architecture had been caught from such carvings of Nature." These architectural qualities made Yellowstone famous and help to account for its popularity ever since.

It was such curiosities, more than the primeval wilderness, that inspired Congress to consecrate the first national park as a "pleasuring ground for the use and enjoyment of the people." The very word "park" suggests the role originally foreseen for places like Yellowstone and Yosemite. Both the private landscape parks of the eighteenth century and the urban public parks of the nineteenth were consciously contrived as places for "congregated human life under glorious and necessarily artificial conditions." Park landscapers from "Capability" Brown and Humphrey Repton to Andrew Jackson Downing and Frederick Law Olmsted did not aim at imitating, much less preserving, nature but at selecting from and improving on it. Unlike natural landscapes, parks were dominated by grassy lawns and groves of big trees free of undergrowth and often contained ornamental structures and ruins, real or make-believe.

Only within the past generation has our emphasis shifted from the tangible enjoyment of spectacular features, natural or artificial, to a more abstract appreciation of untouched wilderness. But if intensive use causes landscape maintenance problems at the Yellowstone and Yosemite campgrounds, visitor impact presents potentially graver hazards in such areas as Rocky Mountain National Park or even the Everglades and the Olympics. Far from "parklike" to begin with, these fragile ecosystems can hardly survive mass recreation; as Dasmann has said, they may have been better off when nobody cared about them. The overcrowded wilderness, like the neglected urban park, is the fruit of distinctions increasingly and unwisely drawn between wild and tame, natural and artificial, countryside and megalopolis.

"Civics as an art," remarked the pioneer city planner Patrick Geddes, has to do "not with imagining an impossible no-place where all is well, but with making the most and the best of each and every place, and especially of the city in which we live." We cannot achieve this by hankering after some distant or exclusive paradise and abandoning all else as beyond hope. It is not only in the wilderness that we escape the tensions of civilized life; man-made environments can also be salubrious. To regard everything used as irretrievably spoiled is, moreover, to relegate the wilderness itself to a museum.

City dwellers have come to believe nature can be found only in unspoiled forests and distant mountains. But nature is in fact all around them—in their own backyards and streets, vacant lots and waterfronts. We need to foster an appreciation of nature in all its guises, humanized as well as wild, near as well as far, intimate as well as grand, transplanted as well as preserved.

The once-in-a-lifetime wilderness experience, the two-week cross-country jaunt, the Sunday jog around the reservoir: these are too rare to sustain a real appreciation of nature, even when supplemented by sumptuous picture books of primeval areas. The aspects of the outdoors that people care about most effectively are those with which they are familiar; to love at a distance is not enough.

6

"Public Parks and Private Lives"

by Ira J. Winn

Associate Professor of Education at San Fernando Valley State College in Los Angeles, Ira J. Winn combines an interest in educational curricula reform with problems of conservation and urbanism—the subject of "Public Parks and Private Lives." He has spent the past three years serving as an advisor on secondary education and educational planning in Brazil and has helped develop civic education programs at UCLA and Columbia University. He has authored several articles on educational reform, the latest of which is "The Education *in Environmental Education" in the* Journal of Environmental Education *(summer, 1970).*

One of the great self-perpetuating myths, which conservationists and wilderness devotees have turned into fervent and desperate belief, is that people are naturally pronature. It is said that once exposed to the beauty and the grandeur of forest and mountain, desert and wild beach, the human being is somehow, almost mystically, drawn and held by the attraction of nature. In effect, the battle for an expanded and improved national park and forest system rests on just this faith: that masses of people, unlike the governor of a rich western state, will not conclude that having seen one redwood, you have seen them all.

But it is not stretching things to say that built into our culture is a fear, perhaps even a hatred, of nature. This fear is reflected in the intense anxiety shown by supposedly worldly people when faced with the stark silence, isolation, and pristine intensity of a wilderness setting. Still more obvious is the indifference or aversion to natural beauty that, apparently, is becoming a characteristic of urban society.

Not too many years ago, I happened to be teaching in a classroom facing one of the ugliest and, in some respects, one of the most typical streets of Los Angeles. It was (and is) a clutter of billboards, telephone and power lines, garishly painted shops, and dirty buildings, which blended into a grotesque blur. There was not a tree in sight to break the shameless monotony. "Look out that window," I told the class one day. "Do you see anything lacking, something wrong with the scene?" They looked a long time before someone ventured, "Not enough parking places."

Now, seven or eight years and many miles later, the props have changed countless times, but the scene is frighteningly similar. In the interval between California and Brazil, I have silently watched drab concrete replace the narrow strip of earth around the expensive home of a friend who can't stand the bother of grass. I have listened to arguments over the values involved in permitting gas stations to take over almost every other street corner, and I remember well the angry talk-stopping rebuttal, "That's the American way!" I have guided many American tourists through Europe, young people for the most part, and I still remember my desperation one rare, vivid day sailing deep in the Norwegian fiords, as more than two-thirds of a group preferred to go below and sit out the two hours drinking, rather than watch the spectacular scenery unfold.

The final straw was added in Brazil. A young woman, a native of Manhattan now living in Rio de Janeiro, recently returned to the high-rise sprawl that is rapidly reducing Copacabana to an oceanside Hollywood and Vine. She was filled with a tale of horror about the few weeks she had spent in the country with relatives—on Long Island, outside New York. "Can you imagine," she asked, "being so far out in the sticks that one is almost overwhelmed by the silence of the night? It was dreadful." She repeated the last several times. "And in the morning you can't even sleep because of the singing of those stupid birds. Oh, I could hardly wait to get back to Copacabana and hear the traffic."

Perhaps these are overly harsh memories. Certainly, among a people habituated to modern comforts it should come as no surprise that the physical world is often taken for granted or actually viewed as the enemy. The wishful thinking of the nature-lorist is possibly but a carryover of another time when man's relations to the natural world were more constant and more true. For what should be evident is that people cannot come to identify with nature unless they have grown up with, and become a part of, it. Lacking wilderness oases in the cities, lacking even a backyard tree for a child to climb, urban man is highly unlikely to develop any widespread appreciation of the mysterious beauty of the forest.

Nevertheless, those who struggle for the outdoor life often seem to forget that the battle will be won or lost where the people are. It is not the evergreens or the wildlife that writes letters, sends congressmen to Washington, forms citizens' committees, plans land use, or sells real estate. Nor for that matter is it even the scattered minority still caught up in small-town life who will cast the deciding vote on the future shape of the environment. It is city people with city-bred ideas who will determine the degree and direction of environmental change and control. They will do it by force of numbers and by the power of the bulldozer, if by nothing else.

Given the population explosion, the speed of modern transportation, and the conglomerate power of impersonal commercial interests, the contemporary conservation battle is only now taking form. It has been hard enough to awaken people to go along with campaigns for park reserves in isolated regions, but it will be infinitely more difficult to get them to stand up against familiar and closely interlocked groups that bring the struggle home to their own city and pocketbook. Thus conservationists must become even more politically active or they will find themselves politically neutralized and recreationally passé. The struggle for protection of the wilderness, for expansion of those great retreats known as the national parks, cannot be separated from the larger contest, now only beginning, over the way of life in our cities. The campaign for wilderness and the quest for privacy are intimately tied to the search for human dignity, which peaks and erupts in every metropolis.

For the most part the cities are creating citizens who are growingly alienated, both from each other and from the natural world. Nature in our modern American cities, cut off from its organic relation to people and to structures, has become merely decorative. Here a vine is put in to help cover another architectural mistake; there some trees are left to shade a sidewalk or as a vestigal break between property lines. But nature as the grand intermediary between man and his work is almost nowhere to be found. At the same time, urban man is faced with a marked diminution in the use of nature as a retreat from the press of the commercial world, and a startling intensification in nervous stimulation from the constant barrage of voices, signals, and raw information. In time this incessant stimulation becomes self-justifying and outwardly self-fulfilling, and he feels nervous and alone if left without his familiar turmoil. As first noted by the German sociologist Georg Simmel at the turn of the century, this phenomenon of metropolitan life forces man constantly to shout and boldly to exaggerate himself simply to remain perceptible—even to himself. Discrimination becomes blunted as the individual develops a protective shield, and communication between people decreases as the means of communication increase. Thereby reduced all the more to a mere unit of consumption in a mass market, the individual loses the significance of his freedom as he becomes increasingly unable to differentiate between things of value and, simply, "things."

Rather than being used as the spiritual refuge they were intended to be, what wooded preserves are left in the large cities have commonly fallen into disuse and misuse; some have even been allowed to decay into virtual lairs for the twisted products of urban pressures. And whatever works toward rationalizing "civilized" encroachments on woodland preserves in the cities, in turn contributes to the alienation and frenzied emptiness that has becomes a way of city life.

What is needed is a functional recreation philosophy for our time, one that can deal with the malaise that results from environmental imbalance in our cities. There are several places to begin, and first among them is the long overdue reform in the recreation curriculum of city schools. Often the urban youngster's only intense contact with recreation is through the hopelessly unrealistic curriculum of competitive sports, which stimulates a hunger for nervous excitation and also implants a long-range tendency toward physical passivity. Contrary to popular opinion and coaches' propaganda, youthful experience on the playing fields does not lead to the active life. Instead, the physical potential of a few youngsters is exploited by the coach, the school, and the community, while the majority are conditioned to the spectator state so lucrative to organized sports-business and a fractional minority of big-name players. For most people the active life somehow never proceeds much beyond settling into an armchair—an indulgence in rear-end muscle flexing while absorbed in the frenzied tallying of scores by 10, 18, or 22 young men.

Physical educators should work to instill functional health and exercise habits in all young people, regardless of ability. But in their desperate search for personal glory and some kind of proof of manhood through the winning team, teachers of physical education appear to forget this goal. The emphasis on the winning team and its corollary, spectator sports, contributes greatly to the life pattern that alternates between hypertension and boredom. Life becomes a never-ending search for thrills and distractions. By the age of 25 or 30, the product of this curriculum is fleshy and dull, a physical creep whose greatest athletic effort is caught in the familiar jibe about pushing oneself away from the table. No amount of rationalizing by persons outraged at any criticism of the sacred cows of popularized recreation can negate the terrible truth of a nation addicted to spectatorism and afflicted by a high incidence of heart disease, tension, and obesity. Americans, observed H. L. Mencken, will go anywhere—sitting down.

The schools must effectively begin to teach lifetime health habits combining physical exercise and quiet, studied mental relaxation and contemplation. The stress must shift to the learning of rhythm skills, relaxation and rejuvenation routines, and individualized developmental and corrective motor-muscular and cardiovascular exercises that have great carry-over value to later life. But the disciplines of yoga and sauna, wilderness hiking and camping, eurythmics, synchronized swimming, cross-country skiing, jogging, and bicycling, for example, should not be taught merely as skills or as a means to social status through the medium of rivalry or self-assertion. For these key arts are the best means of sublimating aggression and learning to appreciate the harmonies of nature by feeling the vitality and joy of rhythmic body movement and physical conditioning for its own sake. Taught as a way of life, rather than as children's games, they form a path to lifelong bodily vigor and sensitivity to the mystery of growth. They open the young mind to love of natural beauty and contentment in beautiful and tranquil surroundings, to mastery of the loneliness of contemplation and disciplined individual expression, and awareness of one's own physical being, secure in self-challenge and respectful amidst the awesome power of nature.

Revitalization of the recreation curriculum involves not just a simple change in the sports orientation of the schools, but demands a much more meaningful relation between the theory of good health and the expected behavior of young people and adults. The needed curriculum requires closing the gap between physical education and biology, with especially strong attention to the lessons of ecology. A successful effort at change therefore implies a basic concern with those key balances required of man if he is to live in tune with nature and at peace with himself and his neighbors. The more natural call for tranquility, harmony, and privacy should replace the hunger for game thrills and nervous stimulation.

American educators would do well to restudy the lessons of *Walden*, the better to learn that the lives of men need not be overflowing with either quiet or noisy desperation. Also, from the perspective of an older civilization seeking a way to technological development without sacrificing its cultural traditions and values, they could profit from the wisdom of its new educational creed. The Ethiopian plan of 1963 defines a key goal of secondary education as the necessity "to develop inner resources of the spirit so that one need not depend all the time on the titillation of the senses from the outside." How unlikely that a statement of that kind would find its way into the declared objectives of an American curriculum.

But schools do not exist in a vacuum, and to become wholly functional even the best of curricula will require a complementary change in the physical community environment. A realistic green-space factor for urban land use—one that puts an end to the modern chicken-coop types of concentrated habitation—must be integrated into law and strictly enforced. The real estate interests, which have so splintered our cities and dehumanized urban living in their lust for the quick sell and the split parcel, will have to be fought and brought under control. Perhaps this can best be accomplished through a system of urban-development grants involving combined federal-state action. But in any case the battle to save the cities can be won only if the citizenry finds wise and spirited leadership dedicated to the concept of urban planning and balanced land use. In effect, natural beauty must be programmed into the urban environment. This means that parks truly must be parks, for in the contemporary idiom parks and playgrounds have been made synonymous; it has become nearly impossible to plan for green space without a golf course or other highly competitive game device to monopolize the area. Thus high-density, residential sections of the city must be particularly well planned for an abundance (not merely an "adequacy") of air and sunlight and for the recreation and rehabilitation value that woodland green space offers to clustered human beings.

Frank Lloyd Wright once claimed that he could remake the jumble that is New York and, on the very same area, build from the ground up a new and more populous skyscraper city set amidst parks and gardens. The challenge need not be a visionary dream, for the task of urban reconstruction must begin with a willingness to face the magnitude of the problem and to depart from traditional forms and dependency on patchwork solutions. Twenty or more entirely new, planned cities of about 500,000 persons each are needed in the immediate future to ease the problems of existing cities while also opening possibilities for the best green space development and reconstruction. This is not as impossible a task as it may at first sound, especially as various examples of new-city and town planning already exist. Further, the long-run costs are small compared to those of the present deficit-ridden cities with their impossible burdens of urban sprawl, pollution, blight, overcrowding, and haphazard localization.

Given the scope of modern construction techniques and the mobility that has come with modern transportation, city growth need not be limited to present sites, which commonly developed as a function of the needs of the old western frontier. Today entirely new urban sites can be chosen, using esthetics among the most basic considerations. In the light of past experience, balanced urban land use suggests fixed minimum lot size and minimum green spacing for homes and apartment buildings as an integral part of the incorporation charter of a city. The guarantee would be reinforced by setting aside large belts of parkland to be known as Greenbelt National Parks running through and around the urban areas. These would be made eternally inviolate from all the rapacious interests that to date have managed to pressure and control city and country administrations and thus frustrate sound urban growth.

In other words, the nationally protected park system that for so long has been thought of in terms of an "outback" must be brought home where the people are. Initially, this will be much easier to plan and implement with respect to new cities, but the concept of urban national parks can be applied equally well to the old. For as the old cities (the present ones) lose population to the carefully planned advantages and attractions of the new, greater opportunities will open up for complete reconstruction and rezoning, including land acquisition for the greenbelts by the National Park Service or a new National Urban Woodlands Commission. However much the cities might then tend to grow toward each other or even blend, the protected greenbelts—and what they imply for people's learned desires for living—will work to prevent the kind of urban bog that is virtually taken for granted in the cities of today.

Out of a balanced urban environment can spring a new kind of urbanite, one who has developed the need for intimacy with nature and who has the capacity to enjoy equally the restless searching, the verve and hurly-burly of city life, and the tranquility that can be gained from country atmosphere. A number of mutually reinforcing and amplifying changes in the very style of life would inevitably begin to occur as a function of successful reform of the recreation curriculum and reconstruction of the physical environment. For one thing, active involvement with the natural environment would stimulate new interests and a marked decline in passive recreation. And as the utter starkness of the city environment was broken, relations between people would become softer and more open.

It is true that the numbers of people coming to the wilderness parks would jump, but the kind of visitors would reflect the healthiness of the change in philosophy: fewer gawkers, screamers, and litter throwers, and many more people seeking intimacy with nature and rejuvenation of body and spirit in primitive surroundings. Indeed, the great national parks, which already are threatened or beset with commercial-ization and overcrowding, should be emptied of virtually all permanent settlement and all the distracting intrusions of resortism. As is already planned in part for the Yosemite Valley, modest facilities would be located in discrete clusters near, but outside, the park peripheries, with trailer-camp visitors and passing tourists bused in and out each day. Thus, public policy would reinforce the general direction of the new recreation movement by preserving the parks as refuges for modern man.

Hope for man's future begins in good part with man's longing for nature—a spirit that is too dear a heritage to be left to wither and die out in the concrete emptiness of the modern megalopolis. Not only does the beauty without reflect what man really values and cherishes, but the shape of the environment greatly affects the life forces within. In this sense the larger struggle for the national parks is both crucial and symbolic. It is a measure of modern man.

7

"Nature and the City"

by August Heckscher

August Heckscher is the Administrator of Parks, Recreation and Cultural Affairs in New York City. Prior to assuming this position, Mr. Heckscher served as Director of the Twentieth Century Fund, as special consultant on the arts to President Kennedy, and in a number of public service positions. He holds degrees from Yale and Harvard and honorary degrees from the Regents of the University of New York State, from C. W. Post College, and from Fairleigh Dickinson, Temple, and Brandeis Universities. Mr. Heckscher is an accomplished author and editor, having served for some years as the chief editorial writer for the New York Herald Tribune.

My feelings for the natural world do not derive from professional experience as a naturalist. I can lay no claim to the role of botanist or zoologist, of horticulturist or ornithologist. I am more, I suppose, a philosopher who came to believe quite long ago that to live in close contact with nature is the surest way to the happiness and wisdom of a full life. This is a very old ideal, a characteristic strain in Western civilization from Socrates to Montesquieu, and to our own Thoreau. It has much to do with the world of natural things, but it goes deeper to affirm that man cannot escape from his own origins, that he must know himself, and that to know himself he must recognize that he is part of the mysterious web of organic and inorganic nature.

When man tries to separate himself from this web he rises not to new heights or a truer wisdom, but ascends at best to the nightmares of reason. Montesquieu speaks somewhere of "celestial thoughts and subterranean conduct." It is a chilling, bitter phrase, but it says something important about man and his fate.

In the past it has perhaps been easiest to attain wisdom and humility in the context of a rural existence. To be close to earth and to the changing seasons, to be in some measure one with growing things, is to be healthily reminded of one's human limitations.

I recall some years ago crossing over into the state of Vermont; at the first crossroads where I stopped to ask the way, I remarked on the fact that it was a beautiful day. The Vermonter looked at me quite a while, then he said, "Yes, we've known a few." The subtle rebuke to my enthusiasm, but even more, the feeling of that man's rootedness in the times and seasons, has always remained with me. The days that he had really known, really lived, had kept him human, as I like to think they had made him wise.

In modern life we have gone to the opposite extreme. The heresy of the age is man's arrogance in putting himself above nature, literally outside of it, as he seeks to control and exploit it for his own uses. Never before has it been possible for man to make over entire environments. But the power to make over is the power to destroy the environment upon which all life—including man's—depends. It is no longer a question of whether or not we shall change the face of the planet. But it is a deep and disconcerting question whether we shall change it so radically, so irreversibly, as to disorganize the processes by which life in all its manifoldness is sustained and renewed. To heedlessly promote or condone the extinction of a growing number of plant and animal species, to begin to make over man's psyche, his body, and, indeed, to tamper with his genes is to stand in a new and terrifying relationship to nature.

A saving instinct may nevertheless be at work. Lewis Mumford in *The Myth of the Machine,* a recent study of profound significance, has suggested that this heresy has in fact developed comparatively recently, and that it may be reversed and outlived. We have read the history of our race, he argues, in terms of toolmaking—we have measured man's progress by his capacity to change outward things. But the reconstruction of history and anthropology appears to lead to another conclusion: that man has always sought self-expression above domination; that he has been primarily the seeker and finder rather than the exploiter. If this is true, then we may yet be rescued by our better selves, and a balance may be established between the appetites of men and the physical limitations of the environment.

We shall be prudent, in any case, to the extent that we preserve nature and study its ways; and the midst of a great city is as appropriate a location as any other in which to do this. As Parks Commissioner I have been keenly aware of the responsibility to see that within the gray city of stone and steel there survives a green city accessible to all the citizens. It is sometimes a discouraging task. One looks about at the vast manifestations of the urban scene, the masses of men, the monstrously piled up structures; one remembers Keats' youthful, jubilant lines:

> To one who has been long in
>   city pent,
> 'Tis very sweet to look into
>   the fair
> And open face of heaven—to
>   breathe a prayer
> Full in the smile of the
>   blue firmament. . . .

Yet Keats, when he felt that release, was only a few miles from the center of the London of his day. He had walked across the heath to find pure air and serene skies. In a far more crowded and difficult environment we must find our own "lair of wavy grass." And we must keep it, notwithstanding the threats that press from all directions.

The parks play many roles in the city of today. They contribute significanctly to its esthestic and architectural dimensions, breaking what must otherwise become a homogenized, dull landscape and individualizing the citizen's environment. To live close to a park is to have one's neighborhood defined; it is significant that almost every open space has its organized, and usually vociferous, citizens' group. The parks also provide a setting for recreation and play. More recently they have become in New York the great stage upon which the common values of the community are celebrated, in such strange ways as "happenings" or in the large gatherings at cultural events. But perhaps most importantly, the parks provide the saving reminder that nature is still part of our lives—or, more correctly, that our lives are a part of nature.

New York is fortunate in the wide spaces that have been left to us by earlier generations. Central Park in Manhattan stands as a quiet miracle at the heart of a city seemingly given over to an expression of man's naked power. Not less extraordinary is the green circle of parks carved out in the Bronx before the northward thrust of the city had made itself fully felt. The city is fortunate, too, in the bays, rivers, and seas that surround it—along with their beaches and their vanishing wetlands, a natural resource that is too often unappreciated.

On a gray and windy day last spring, during my first weeks as Parks Commissioner, I sailed out in the motor launch that is the flagship of the Department's navy to explore the bird-populated islands of Jamaica Bay. Another day I stood in a roaring northeaster upon the dramatic outcropping of rock at the point of Pelham Bay Park. On Staten Island, I walked in a light rain through the High Rock nature preserve watching a group of children who had come out of the classroom to learn something of the mystery of a pond's ecology. These experiences represent, for me, some of the more rewarding contacts with the city's natural environment, because these shoreline areas remain relatively unaltered by the hand of man.

And then, on another memorable evening, Mayor Lindsay and I landed by helicopter on the beach at Rockaway. We listened as one of our good park workers explained how our highly prized "beach sanitizer" cleaned the shore so well of all debris, including seashells, that the sands blew away and had to be restored by bringing in a bulldozer. There, indeed, was an example of our ineptitude in attempting to improve upon nature with "management."

A surprising number of natural scenes remain in New York, but the city presses inexorably upon all the open areas. The road builders have already cut across some of our greatest parks, and threaten others. Institutional buildings within or alongside the parks seek, as if by a natural law, to expand. Pollution of earth, air, and water limits the usefulness of the parks and shorefronts. In a frantic search for places to dispose of its ever-mounting wastes, the modern city looks to parks as an inviting, though temporary, expedient.

A few victories, a few hoped for reversals of the current trends, give meaning to the work of a city Parks Commissioner. Soon after I was appointed, a beautiful wetland in the Split Rock section of Pelham Bay Park was slated, under a long-standing plan, to become a dumping place for raw garbage. A million years had gone into the making of that rich, mysterious piece of earth and water; if it were touched now, another million years could not undo the damage. The Mayor, the head of the City Planning Commission, the Commissioner of Sanitation, and I gathered one late spring evening on the spot. The astonishing but highly gratifying outcome of our meeting was that Split Rock should be saved!

The people of the Bronx continue to generate the wastes that are part of contemporary civilization, and we still have not resolved the problem of how these are to be disposed of in the interval before long-range plans can be put into effect. But the wetland is, for now, secure. And soon children from the whole city will be going to a simple nature center there, where they can see and learn of nature's strange workings, and explore a wealth such as few cities possess.

Jamaica Bay is kept as a bird sanctuary, in spite of pressures to give over its tide-washed islands to housing, industry, or mass recreation. The jets of nearby Kennedy Airport have not driven off nature's own fliers, and if the city takes the sensible step of decentralizing its air traffic, birds may still be at home there a generation from now, perhaps several generations hence. The green belt on Staten Island persists precariously, while authorities in Albany and Washington seem unwilling to give final acknowledgment to the supremacy of man's stake in nature over his disposition to build eight-lane expressways. At the tip of the Rockaways, Breezy Point displays the skeletons of half-completed housing projects, halted by court order while men exhibit the good sense to weigh the merits of natural beauty over one more extension of concrete.

Last year, when the poet laureate of England died, it was good to recall his feeling for those small, cherished parts of earth that contain much that is important to man's wholeness and sanity. "Some places," John Masefield had written, "seem to have a memory and a living spirit. We do well to keep such places holy, and to seek what they have to tell us."

In the modern, all-consuming city such locations may be those that mark the journey of a good man or note some other moment in history. But most often they are simple bits of wildness left for us to cherish and keep. What these places have to tell us is important beyond the power of words to convey. On our capacity to listen to the lesson they teach, and truly to understand it, may well depend man's survival on this planet.

8

"The Myth of the Peaceful Atom"

by Richard Curtis and Elizabeth Hogan

The research leading to the publication of "The Myth of the Peaceful Atom" was begun in 1963 by Elizabeth Hogan. Co-author Richard Curtis joined in the effort in 1967 and the two have since conducted a thorough study of the ramifications of the commercial use of atomic energy. Their investigation led to consultations with nuclear scientists, engineers, and biologists, and has culminated in a book on peaceful uses of the atom, which was published by Doubleday in 1969. Both Hogan and Curtis are free-lance writers who work and live in New York City.

"What is past is past, and the damage we may already have done to future generations cannot be rescinded, but we cannot shirk the compelling responsibility to determine if the course we are following is one we should be following."

So said Senator Thruston B. Morton of Kentucky on February 29, 1968, upon introducing into Congress a resolution calling for comprehensive review of federal participation in the atomic energy power program. Admitting he had been remiss in informing himself on this "grave danger," Morton said he had now looked more deeply into nuclear power safety and was "dismayed at some of the things I have found—warnings and facts from highly qualified people who firmly believe that we have moved too fast and without proper safeguards into an atomic power age."

Senator Morton's resolution on nuclear power was by no means the only one before Congress in 1968. Indeed, more than two dozen legislators urged investigation and reevaluation of this program. This fact may come as a surprise to much of the public, for the belief is widespread that the nuclear reactors being built to generate electricity for our cities are safe, reliable, and pollution-free. But a rapidly growing number of physicists, biologists, engineers, public health officials, and even staff members of the Atomic Energy Commission itself—the government bureau responsible for regulation of this force—have been expressing serious misgivings about the planned proliferation of nuclear power plants. In fact, some have indicated that nuclear power, which Supreme Court Justices William O. Douglas and Hugo L. Black described as "the most deadly, the most dangerous process that man has ever conceived," represents the gravest pollution threat yet to our environment.

As of June, 1968, 15 commercial nuclear power plants were operating or operable within the United States, producing about one per cent of our current electrical output. The government, however, has been promoting a plan by which 25 per cent of our electric power will be generated by the atom by 1980, and half by the year 2000. To meet this goal, 87 more plants are under construction or on the drawing boards. Although atomic power and reactor technology are still imperfect sciences, saturated with hazards and unknowns, these reactors are going up in close proximity to heavy population concentrations. Most of them will be of a size never previously attempted by scientists and engineers. They are, in effect, gigantic nuclear experiments.

As most readers will recall, atomic reactors are designed to use the tremendous heat generated by splitting atoms. They are fueled with a concentrated form of uranium stored in thin-walled tubes bound together to form subassemblies. These are placed in the reactor's core, separated by control rods that absorb neutrons and thus help regulate chain reactions of splitting atoms. When the rods are withdrawn, the chain reactions intensify, producing enormous quantities of heat. Coolant circulated through the fuel elements in the reactor core carries the heat away to heat-exchange systems, where water is brought to a boil. The resultant steam is employed to turn electricity-generating turbines.

Stated in this condensed fashion, the process sounds innocuous enough. Unfortunately, however, heat is not the only form of energy produced by atomic fission. Another is radioactivity. During the course of operation, the fuel assemblies and other components in the reactor's core become intensely radioactive. Some of the fission by-products have been described as a million to a billion times more toxic than any known industrial chemical. Some 200 radioactive isotopes are produced as by-products of reactor operation, and the amount of just one of them, strontium-90, accumulated in a reactor of even modest (100-200 megawatt) size, after it has been operative for six months, is equal to what would be produced by the explosion of a bomb 190 times more powerful than the one dropped on Hiroshima.

Huge concentrations of radioactive material are also to be found in nuclear fuel-reprocessing plants. Because the intense radioactivity in a reactor core eventually interferes with the fuel's efficiency, the spent fuel assemblies must be removed from time to time and replaced by new, uncontaminated ones. The old ones are transported to reprocessing plants where the contaminants are separated from the salvageable fuel as well as from plutonium, a valuable by-product. Since no satisfactory means have been found for neutralizing or for safely releasing into the environment the radioactive liquid containing the contaminants, it must be stored until it is no longer dangerous. Thus, reprocessing plants and storage areas are immense repositories of "hot" and "dirty" material. Furthermore, routes between nuclear power plants and the reprocessing facility carry traffic bearing high quantities of such material.

Even from this glimpse it will be apparent that public and environmental safety depend on the flawless containment of radioactivity every step of the way. For, owing to the incredible potency of fission products, even the slightest leakage is harmful and a massive release would be catastrophic. The fundamental question, then, is how heavily can we rely on human wisdom, care, and engineering to hold this peril under absolute control?

Abundant evidence points to the conclusion that we cannot rely on it at all.

The hazards of peaceful atomic power fall into two broad categories: the threat of violent, massive releases of radioactivity or that of slow, but deadly, seepage of harmful products into the environment.

Nuclear physicists assure us that reactors cannot explode like atomic bombs because the complex apparatus for detonating an atomic warhead is absent. This fact, however, is of little consolation when it is realized that only a *conventional* explosion, which ruptures the reactor mechanism and its containment structure, could produce havoc on a scale eclipsing any industrial accident on record or any single act of war, including the atomic destruction of Hiroshima or Nagasaki.

There are numerous ways in which such an explosion can take place in a reactor. For example, liquid sodium, which is used in some reactors as a coolant, is a devilishly tricky element that under certain circumstances burns violently on contact with air. Accidental exposure of sodium could initiate a chain of reactions: rupturing fuel assemblies, damaging components and shielding, and destroying primary and secondary emergency safeguards. If coolant is lost, as it could be in some types of reactors, fuel could melt and recongeal, forming "puddles" that could explode upon reaching a critical size. If these explosions are forceful enough, and safeguards fail, some of the fission products could be released outside the plant and into the environment in the form of a gas or a cloud of fine radioactive particles. Under not uncommon atmospheric conditions such as an "inversion," in which a layer of warm air keeps a cooler layer from rising, a blanket of radioactivity could spread insidiously over the countryside. Another possibility is that fission products could be carried out of the reactor and into a city's watershed, for all reactors are being built on lakes, rivers, or other bodies of water for cooling purposes.

What would be the toll of such a calamity?

In 1957 the Atomic Energy Commission issued a study (designated Wash. —740), largely prepared by the Brookhaven National Laboratory, that attempted to assess the probabilities of such "incidents" and the potential consequences. Some of its findings were stupefying: From the explosion of a 100-200 megawatt reactor, as many as 3,400 people could be killed, 43,000 injured, and as much as 7 billion dollars of property damage done. People could be killed at distances up to 15 miles and injured up to 45. Land contamination could extend for far greater distances: agricultural quarantines might prevail over an area of 150,000 square miles, more than the combined areas of Pennsylvania, New York, and New Jersey.

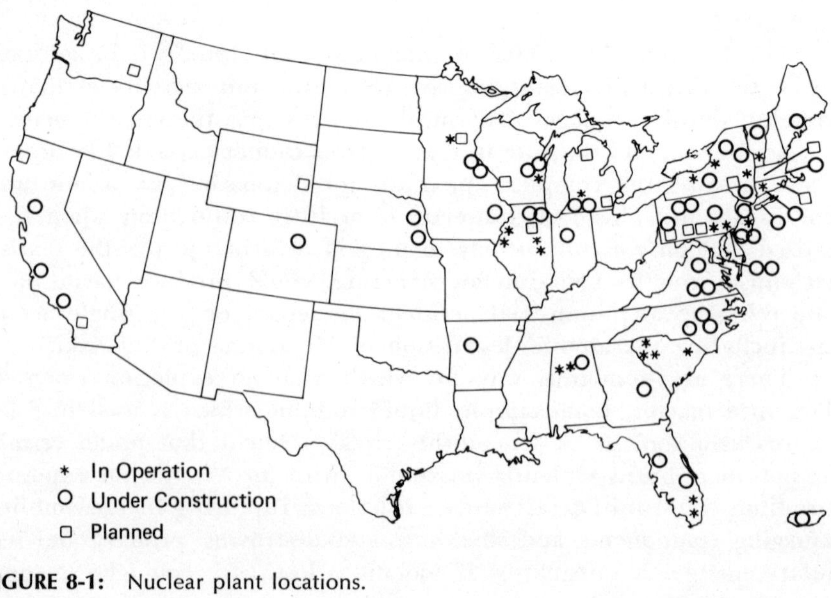

FIGURE 8-1: Nuclear plant locations.

The awful significance of these figures is difficult to comprehend. By way of comparison, we might look at one of the worst industrial accidents of modern times: the Texas City disaster of 1947 when a ship loaded with ammonium nitrate fertilizer exploded, virtually leveling the city, killing 561 people, and causing an estimated $67 million worth of damage. Appalling as this catastrophe was, however, it does not begin to approach the potential havoc that would be wreaked by a nuclear explosion occurring in one of the plants now being constructed close to several American cities.

The scientists and engineers who produced the Brookhaven Report optimistically ventured to give high odds against such an occurrence, asserting that the structures, systems, and safeguards of atomic plants were so engineered as to render it practically incredible. At the same time, though, the report was replete with such statements as:

> "The cumulative effect of radiation on physical and chemical properties of materials, after long periods of time, is largely unknown."
>
> "Much remains to be learned about the characteristics and behavior of nuclear systems."
>
> "It is important to recognize that the magnitudes of many of the crucial factors in this study are not quantitatively established, either by theoretical and experimental data or adequate experience."

Even if the report had been founded on more substantial understanding of natural and technical processes, many of the grounds on which the Brookhaven team based its conclusions are shaky at best.

For one thing, all of us are familiar with technological disasters that have occurred against fantastically high odds: the sinking of the "unsinkable" *Titanic,* or the November 9, 1965, "blackout" of the northeastern United States, for example. The latter happening illustrates how an "incredible" event can occur in the electric utility field, most experts agreeing that the chain of circumstances that brought it about was so improbable that the odds against it defy calculation.

Congressional testimony given in 1967 by Dr. David Okrent, a former chairman of the AEC's Advisory Committee on Reactor Safeguards, demonstrated that fate is not always a respecter of enormously adverse odds. "We do have on record cases where, for example, an applicant, appearing before an atomic safety and licensing board, stated that a mathematical impossibility had occurred; namely, one tornado took out five separate power lines to a reactor. If one calculated strictly on the basis of probability and multiplied the probability for one line five times, you get a very small number indeed," said Dr. Okrent, "but it happened."

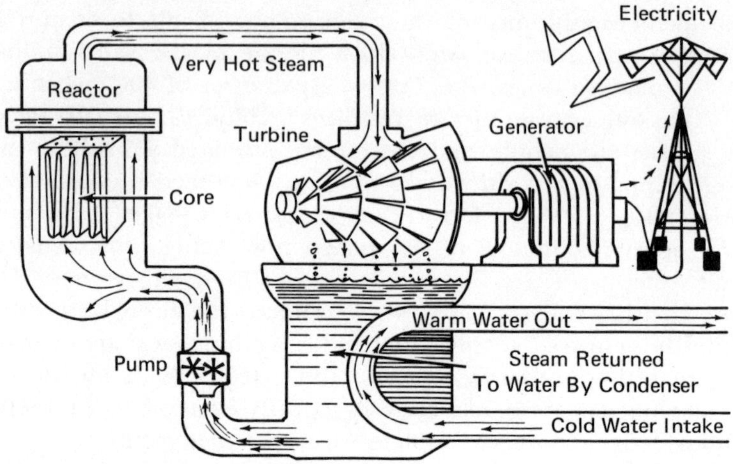

FIGURE 8-2: Boiling water reactor.

A disturbing number of reactor accidents have occurred—with sheer luck playing an important part in averting catastrophe—that seem to have been the product of incredible coincidences. On October 10, 1957, for instance, the Number One Pile (reactor) at the Windscale Works in England malfunctioned, spewing fission products over so much territory that authorities had to seize all milk and growing foodstuffs in a 400-square-mile area around the plant. A British report on the incident stated that *all* of the reactor's containment features had failed. And, closer to home, a meltdown of fuel in the Fermi reactor in Lagoona Beach, Michigan, in October, 1966, came within an ace of turning into a nuclear "runaway." An explosive release of radioactive materials was averted, but the failures of Fermi's safeguards made the event, in the words of Sheldon Novick in *Scientist and Citizen,* "a bit worse than the 'maximum credible accident.' "

The atomic industry has attempted to design components and safeguards so that failure of one vital system in a plant will not affect another, resulting in a "house of cards" collapse. However, two highly regarded authorities, Theos J. Thompson and J. G. Beckerley, in a book on reactor safety advise us not to place too much faith in claims of independent safeguards: "A structure as complex as a reactor and involving as many phenomena is likely to have relatively few completely independent components." Many manufacturers and utility operators have resisted the idea of producing "redundant safeguards" on the grounds of excessive cost.

Investigations of reactor breakdowns usually disclose a number of small, seemingly unrelated failures, which snowballed into one big one.

A design flaw or a human error, a component failure here, an instrumentation failure there—all may coincide to contribute to the total event. Thompson and Beckerley, examining several atomic plant accidents, pinpointed 13 different contributing causes in three of the accidents that had occurred up to the time of their 1964 study.

Among the many factors contributing to reactor accidents, the human element is the most difficult to quantify. And perhaps for that reason, it has been largely overlooked in the AEC's assessments of reactor safety. Yet, a private researcher of nuclear accidents, Dr. Donald Oken, M.D., Associate Director of the Psychosomatic and Psychiatric Institute of Michael Reese Hospital in Chicago reported: "A review of reports of past criticality and reactor incidents and discussions held with some of the health personnel in charge reveal a number of striking peculiarities in the behavior of many of those involved—in which they almost literally asked for trouble."

AEC annuals are full of reports of human negligence: 3,844 pounds of uranium hexafluoride lost owing to an error in opening a cylinder; a $220,000 fire in a reactor because of accidental tripping of valves by electricians during previous maintenance work: numerous vehicular accidents involving transport of nuclear materials. None of these accidents led to disaster, but who will warrant that, with the projected proliferation of power plants and satellite industries in the coming decade, a moment's misjudgment will not trigger a nightmare? Perhaps worse, the likelihood of sabotage has scarcely been weighed, despite a number of incidents and threats.

It should be apparent that if men are to build safe, successful reactors, the whole level of industrial workmanship, engineering, inspection, and quality control must be raised well above prevailing levels. The more sophisticated the technology, the more precise the correspondence between the subtlest gradations of care or negligence and that technology's success or failure. When meters, grams, and seconds are no longer good enough, and specifications call for millimeters, milligrams, and milliseconds, the demands made on men, material, and machinery are accordingly intensified. Minute lapses that might be tolerable in a conventional industrial procedure will wreck the more exacting one. And when the technology is not only exacting but hazardous in the extreme, then a trivial oversight, a minor defect, a moment's inattention may spell doom.

While there is little doubt that American technology is the most refined on earth, there is ample reason to believe that it has more than met its match in the seemingly insurmountable problems posed by the peaceful atom. Societies of professional engineers, and others concerned with establishing technical and safety criteria for the nuclear industry, have described between 2,800 and 5,000 technical standards that are necessary for a typical reactor power plant in such areas as materials, testing, design, electrical gear, instrumentation, plant equipment, and processes. Yet, due to the rapidity with which the nuclear industry has developed, as of March, 1967, only about 100 of these had been passed on and approved for use.

It is not surprising, then, to learn that serious technical difficulties are turning up in reactor after reactor. At the Big Rock Point Nuclear Plant, a relatively small reactor near Charlevoix, Michigan, control rods were found sticking in position, studs failing or cracked, screws jostled out of place and into key mechanisms, a valve malfunctioning for more than a dozen reasons, foreign material lodging in critical moving parts, and welds cracked on every one of sixteen screws holding two components in place. A reactor at Humboldt Bay in California manifested cracks in the tubes containing fuel: in order to keep costs down, stainless steel had been used instead of a more reliable alloy. The Oyster Creek plant in New Jersey showed cracks in 123 of 137 fuel tubes, and welding defects at every point where tubes and control-rod housings were joined around the reactor's vessel. Reactors in Wisconsin, Minnesota, Connecticut, Puerto Rico, New York, and elsewhere have experienced innumerable operating difficulties, and some, such as the $55 million Hallam plant in Nebraska have been forced to shut down for good, owing to plant malfunction.

Chilling parallels can be drawn between failures in nuclear utility technology and in the nuclear submarine program. In October, 1962, Vice Admiral Hyman G. Rickover, Director of AEC's Division of Naval Reactors, took the atomic industry to task in a speech in New York City.

"It is not well enough understood that conventional components of advanced systems must nessarily meet higher standards. Yet it should be obvious that failures that would be trivial if they occurred in a conventional application will have serious consequences in a nuclear plant because here radioactivity is involved. . . ."

Rickover went on to cite defective welds, forging materials substituted without authorization, violations of official specifications, poor inspection techniques, small and seemingly "unimportant" parts left out of components, faulty brazing of wires, and more. "I assure you," he declared, "I am not exaggerating the situation; in fact, I have understated it. For every case I have given, I could cite a dozen more."

The following April, the U.S. atomic submarine *Thresher,* while undergoing a deep test dive some 200 miles off the Cape Cod coast, went down with 112 naval personnel and 17 civilians and never came up again. Subsequent investigation revealed that the sub suffered from many of the same ailments described in Rickover's speech. "It is extremely unfortunate," said Senator John O. Pastore, chairman of the joint congressional committee that held hearings on the disaster, "that this tragedy had to occur to bring a number of unsatisfactory conditions into the open." We must now ask if the same will one day be said about a power plant near one of our large cities.

If a major reactor catastrophe did occur there is good reason to believe that the consequences would be far worse than even the dismaying toll suggested by the 1957 Brookhaven Report, for a number of developments since then have made the threat considerably more formidable.

The Brookhaven Report's accident statistics, for instance, pertained to a reactor of between 100 and 200 megawatts. But while the 15 reactors currently operating in the United States average about 186 megawatts, the 87 plants going up or planned for the next decade are many times that size. Thirty-one under construction average about 726 megawatts; 42 in the planning stage average 832; 14 more, planned but without reactors ordered, will average 904. Some, such as those slated for Illinois, California, Alabama, and New York anticipate capacities of more than 1,000 megawatts. Con Edison has just announced it intends to build four units of 1,000 megawatts each on Long Island Sound near New Rochelle in teeming Westchester County—four nuclear reactors, each with a capacity five to ten times that of the reactor described in the Brookhaven Report.

These facilities will accordingly contain more uranium fuel, and because it is costly to replace spent fuel assemblies (this delicate and dangerous process can take six weeks or longer), the new reactors are designed to operate without fuel replacement far beyond the six months posited in the Brookhaven Report. As a result, the buildup of toxic fission products in tomorrow's reactors will be far greater than at present, and an accident occurring close to the end of the "fuel cycle" in such a plant could release fantastic amounts of radioactive material.

Most serious of all, perhaps, is that tomorrow's reactors are now slated for location in close proximity to population concentrations. While the Brookhaven Report had its hypothetical reactor situated about 30 miles from a major city, many of tomorrow's atomic plants will be much closer. Although the AEC has drafted "guidelines" for siting reactors, the Commission has failed to make utilities adhere to them. In 1967, Clifford K. Beck, AEC's Deputy Director of Regulation, admitted to the Joint Committee on Atomic Energy that nuclear plants in Connecticut, California, New York, and other locations "have been approved with lower distances than our general guides would have indicated when they were approved."

Also, we must remember that while a reactor may not be near the legal boundaries of a metropolis, it may lie close to a population center. Thus, while Con Edison's Indian Point plant is 24 miles from New York City (two more plants are now being built there), it is within 10 miles of an estimated population of 155,510. It need only be recalled that the Brookhaven Report foresaw people being killed by a major radioactive release at distances up to 15 miles to realize the significance of these figures.

In a recent study of nuclear plant siting made by W. K. Davis and J. E. Robb of San Francisco's Bechtel Corporation, the locations of 42 nuclear power plants (some proposed, some now operable) were examined with respect to population centers inhabited by 25,000 residents or more. Their findings are unnerving: only *two* plants in operation or planned are more than 30 miles from a population center. Of the rest, 14 are between 20 and 27 miles away, 15 between 10 and 16 miles, and 11 between 1 and 9 miles.

Is it necessary to build atomic plants so big and so close? The answer has to do with economics. The larger a facility is, the lower the unit cost of construction and operation and the cheaper the electricity. The longer the fuel cycle, the fewer the expensive shutdowns while spent fuel assemblies are replaced. The closer the plant is to the consumer, the lower the cost of rights of way, power lines, and other transmission equipment.

On a few occasions an aroused public has successfully opposed the situation of plants near population centers. When the Pacific Gas and Electric Company persisted in trying to build a reactor squarely over earthquake faults in an area of known seismic activity—the site was Bodega Head, north of San Francisco—a courageous conservation group forced the company to back down. It has been suggested, though, that the group might not have won had not the Alaskan earthquake of 1964, occurring while the fight was going on, underscored the recklessness of the utility's scheme.

Announcement by Con Edison at the end of 1962 of its proposal to build a large nuclear plant in Ravenswood, Queens, close to the center of New York City brought a storm of frightened and angry protest. Although the utility's chairman noted, "We are confident that a nuclear plant can be built in Long Island City, or in Times Square for that matter, without hazard to our own employees working in the plant or to the community," David E. Lilienthal, the former head of the AEC, had a contrary opinion, declaring he "would not dream of living in Queens if a huge nuclear plant were located there." Outraged citizens and a number of noted scientists prevailed.

For the most part, however, the battle has been a losing one. Con Edison, for example, after its defeat in the Ravenswood fight, has just announced an interest in building a reactor on Welfare Island, literally a stone's throw from midtown Manhattan. Also, New York's Governor Nelson Rockefeller has gone on record advocating an $8 billion electric power expansion program based extensively on nuclear energy. The state legislature approved of the program, and in 1968, voted to bolster the plan with state subsidies.

Some of the deepest concern about the size and location of atomic plants has been expressed by members of the AEC themselves. "The actual experience with reactors in general is still quite limited," said Dr. Clifford K. Beck, AEC's Deputy Director of Regulation, in 1967 congressional hearings, "and with large reactors of the type now being considered, it is nonexistent. Therefore, because there would be a large number of people close by and because of lack of experience, it is . . . a matter of judgment and prudence at present to locate reactors where the protection of distance will be present."

Beck's statement is mild compared to that made in the same hearings by Nunzio J. Palladino, Chairman of the AEC's Advisory Committee on Reactor Safeguards for 1967, and Dr. David Okrent, former Chairman for 1966: "the ACRS believes that placing large nuclear reactors close to population centers will require considerable further improvements in safety, and that *none of the large power reactors now under construction is considered suitable for location in metropolitan areas* [our italics]."

The threat of a nuclear plant catastrophe constitutes only half of the double jeopardy in which atomic power has placed us. For even if no such calamity occurs, the gradual exhaustion of what one scientist terms our environmental "radiation budget," due to unavoidable releases of radioactivity during normal operation of nuclear facilities, poses an equal and possibly more insidious threat to all living things on earth.

Most of the fission products created in a reactor are trapped. Contaminated solids, liquids, and gases are isolated, allowed to decay for a short period of time, then concentrated and shipped in drums to storage areas. These are called "high-level wastes." But technology for retaining all radioactive contaminants, is either unperfected or costly, and much material of low-level radioactivity is routinely released into the air or water at the reactor site. These releases are undertaken in such a way, we are told, as to insure dispersion or dilution sufficient to prevent any predictable human exposure above harmful levels. Thus, when atomic power advocates are asked about the dangers of contaminating the environment, they imply that the relatively small amounts of radioactive materials released under "planned" conditions are harmless.

This view is a myth.

In the first place, many waste radionuclides take an extraordinarily long time to decay. The half-life (the time it takes for half of an element's atoms to disintegrate through fission) of strontium-90, for instance, is more than 27 years. Thus, even though certain long-lived isotopes are widely dispersed in air or diluted in water, their radioactivity does not cease. It remains, and over a period of time accumulates. It is therefore not pertinent to talk about the safety of any single release of "hot" effluents into the environment. At issue, rather, is their duration and cumulative radioactivity.

Further, many radioactive elements taken into the body tend to build up in specific tissues and organs to which those isotopes are attracted, increasing by many times the exposure dosage in those local areas of the body. Iodine-131, for instance, seeks the thyroid gland; strontium-90 collects in the bones; cesium-137 accumulates in muscle. Many isotopes have long half-lives, some measurable in decades.

Two more factors controvert the view that carefully monitored releases of low-level radioactivity into the environment are not pernicious. First, there is apparently no radiation threshold below which harm is impossible. Any dose, however small, will take its toll of cell material, and that damage is irreversible. Second, it may take decades for organic damage, or generations for genetic damage, to manifest itself. In 1955, for example, two British doctors reported a case of skin cancer—ultimately fatal—that had taken forty-nine years to develop following fluoroscopic irradiation of a patient.

Still another problem has received inadequate attention. Man is by no means the only creature in whom radioactive isotopes concentrate. The dietary needs of all plant and animal life dictate intake of specific elements. These concentrate even in the lowest and most basic forms of life. They are then passed up food chains, from grass to cattle to milk to man, for example. As they progress up these chains, the concentrations often increase, sometimes by hundreds of thousands of times. And if these elements are radioactive. . . .

Take zinc-65, produced in a reactor when atomic particles interact with zinc in certain components. Scrutiny of the wildlife in a pond receiving runoff from the Savannah River Plant near Aiken, South Carolina, disclosed that while the water in that pond contained only infinitesimal traces of radioactive zinc-65, the algae that lived on the water had concentrated the isotope by nearly 6,000 times. The bones of bluegills, an omnivorous fish that feeds both on algae and on algae-eating fish, showed concentrations more than 8,200 times higher than the amount found in the water. Study of the Columbia River, on which the Hanford, Washington, reactor is located, revealed that while the radioactivity of the water was relatively insignificant: 1. the radioactivity of the river plankton was 2,000 times greater; 2. the radioactivity of the fish and ducks feeding on the plankton was 15,000 and 40,000 times greater, respectively; 3. the radioactivity of young swallows fed on insects caught by their parents near the river was 500,000 times greater; and 4. the radioactivity of the egg yolks of water birds was more than a million times greater.

Here then are clear illustrations of the ways in which almost undetectable traces of radioactivity in air, water, or soil may be progressively concentrated, so that by the time it ends up on man's plate or in his glass it is a tidy package of poison.

That nuclear facilities are producing dangerous buildups of radioisotopes in our environment can be amply documented. University of Nevada investigators, seeking a cause for concentrations of iodine-131 in cattle thyroids in wide areas of the western United States, concluded that "the principal known source of I-131 that could contribute to this level is exhaust gases from nuclear reactors and associated fuel-processing plants."

In his keynote address to the Health Physics Society Symposium at Atlanta, Georgia, early in 1968, AEC Commissioner Wilfred E. Johnson admitted that the release into the atmosphere of tritium and noble gases such as krypton-85 would present a potential problem in the future, and that, as yet, scientists had not devised a way of solving it. Krypton-85, although inert, has a 10-year half-life and tends to dissolve in fatty tissue, meaning fairly even distribution throughout the human body. Krypton-85 is particularly difficult to filter out of reactor discharges, and the accumulation of this element alone may exhaust as much as two-thirds of the "average" human's "radiation budget" for the coming century, based on the standards established by the National Committee on Radiation Protection and Measurement.

That "low-level" waste is a grossly deceptive term is obvious. In his book *Living with the Atom,* author Ritchie Calder in 1962 described an "audit" of environmental radiation that he and his colleagues, meeting at a symposium in Chicago, drew up to assess then current and future amounts of radioactivity released into atmosphere and water. Speculations covered the period 1955-65, and because atomic power plants were few and small during that time, the figures are more significant in relation to the future. Tallying "planned releases" of radiation from such sources as commerical and test reactors, nuclear ships, uranium mills, plutonium factories, and fuel-reprocessing plants, Calder's group came to a most disquieting conclusion: "By the time we had added up all the curies which might predictably be released, by all those peaceful uses, into the environment, it came to about 13 million curies per annum." A "curie" is a standard unit of radioactivity whose lethality can be appreciated from the fact that one trillionth of one curie of radioactive gas per cubic meter of air in a uranium mine is ten times higher than the official maximum permissible dose.

Calder's figures did not include fallout due to bomb testing and similar experiments, nor did they take into account possible reactor or nuclear transportation accidents. Above all, they did not include possible escape of stored high-level radioactive wastes, the implications of which were awesome to contemplate: "what kept nagging us was the question of waste disposal and of the remaining radioactivity which must not get loose. We were told that the dangerous waste, which is kept in storage, amounted to 10,000 million curies. If you wanted to play 'the number's game' as an irresponsible exercise, you could divide this by the population of the world and find that it is over 3 curies for every individual."

Exactly what does Calder mean by "the question of waste disposal"?

It has been estimated that a ton of spent fuel in reprocessing will produce from forty to several hundred gallons of waste. This substance is a violently lethal mixture of short- and long-lived isotopes. It would take five cubic miles of water to dilute the waste from just *one* ton of fuel to a safe concentration. Or, if we permitted it to decay naturally until it reached the safe level—and the word "safe" is used advisedly—just one of the isotopes, strontium-90, would still be damaging to life 1,000 years from now, when it will have only one seventeen-billionth of its current potency.

There is no known way to reduce the toxicity of these isotopes; they must decay naturally, meaning *virtually perpetual containment.* Unfortunately, mankind has exhibited little skill in perpetual creations, and procedures for handling radioactive wastes leave everything to be desired. Formerly dumped in the ocean, the most common practice today is to store the concentrates in large steel tanks shielded by earth and concrete. This method has been employed for some twenty years, and about 80 million gallons of waste are now in storage in about 200 tanks. This "liquor" generates so much heat it boils by itself for years. Most of the inventory in these caldrons is waste from weapons production, but within thirty years, the accumulation from commercial nuclear power will soar if we embark upon the expansion program now being promoted by the AEC. Dr. Donald R. Chadwick, chief of the Division of Health of the U.S. Public Health Service, estimated in 1963 that the accumulated volume of waste material would come to two billion gallons by 1995.

It is not just the volume that fills one with sickening apprehension but the techniques of disposing of this material. David Lilienthal put his finger on the crux of the matter when he stated: "These huge quantities of radioactive wastes must somehow be removed from the reactors, must —without mishap—be put into containers that will never rupture; then these vast quantities of poisonous stuff must be moved either to a burial ground or to reprocessing and concentration plants, handled again, and disposed of, by burial or otherwise, with a risk of human error at every step." Nor can it be stressed strongly enough that we are not discussing a brief danger period of days, months, or years. We are talking of periods "longer," in the words of AEC Commissioner Wilfred E. Johnson, "than the history of most governments that the world has seen."

Yet already there are many instances of the failure of storage facilities. An article in an AEC publication has cited nine cases of tank failure out of 183 tanks located in Washington, South Carolina, and Idaho. And a passage in the AEC's authorizing legislation for 1968 called for funding $2,500,000 for the replacement of failed and failing tanks in Richland, Washington. "There is no assurance," concluded the passage, "that the need for new waste storage tanks can be forestalled." If this is the case after twenty years of storage experience, it is beyond belief that this burden will be borne without some storage failures for centuries in the future. Remember too, that these waste-holding "tank farms" are vulnerable to natural catastrophes such as earthquakes, and to man-made ones such as sabotage.

Efforts are of course being made toward effective handling of the waste problems, but many technical barriers must still be overcome. It is unlikely they will all be overcome by the end of the century, when waste tanks will boil with 6 billion curies of strontium-90, 5.3 billion curies of cesium-137, 6.07 billion curies of prometheum-147, 10.1 billion curies of cerium-144, and millions of curies of other isotypes. The amount of strontium-90 alone is 30 times more than would be released by the nuclear war envisioned in a 1959 congressional hearing.

The burden that radioactive wastes place on future generations is cruel and may prove intolerable. Physicist Joel A. Snow stated it well when he wrote in *Scientist and Citizen:* "Over periods of hundreds of years it is impossible to ensure that society will remain responsive to the problems created by the legacy of nuclear waste which we have left behind."

"Legacy" is indeed a gracious way of describing the reality of this situation, for at the very least we are saddling our children and their descendants with perpetual custodianship of our atomic refuse, and at worst may be dooming them to the same agonizing afflictions and deaths suffered by those who survived Hiroshima. Radiation has been positively linked to cancer, leukemia, brain damage, infant mortality, cataracts, sterility, genetic defects and mutations, and general shortening of life.

The implications for the survival of mankind can be glimpsed by considering just one of these effects, the genetic. In a 1960 article, James F. Crow, Professor of Genetics at the University of Wisconsin School of Medicine and president of the Genetics Society of America, stated that for every roentgen of slow radiation—the kind we can expect to receive in increasing doses from peacetime nuclear activity—about five mutations will result per 100 million genes exposed, meaning that "after a number of generations of exposure to one roentgen per generation, about one in 8,000 . . . in each generation would have severe genetic defects attributable to the radiation."

The Atomic Energy Commission is aware of the many objections that have been raised to the atomic power program: why does it continue to encourage it? Unfortunately, the Commission must perform two conflicting roles. On the one hand, it is responsible for regulating the atomic power industry. But on the other, it has been charged by Congress to promote the use of nuclear energy by the utility industry. Because of its involvement in the highest priorities of national security, enormous power and legislative advantages have been vested in the AEC, enabling it to fulfill its role as promoter with almost unhampered success—while its effectiveness as regulator has gradually atrophied. The Commission consistently denies claims that atomic power is heading for troubled waters, optimistically reassuring critics that these plants are safe, clean neighbors.

The fact that there is no foundation for this optimism is emphasized by the insurance situation on atomic facilities. Despite the AEC's own assertion that as much as $7 billion in property damage could result from an atomic power plant catastrophe, the insurance industry, working through two pools, will put up no more than $74 million, or about one per cent, to indemnify equipment manufacturers and utility operators against damage suits from the public. The federal government will add up to $486 million more, but this still leaves more than $6 billion in property damages to be picked up by victims of a Brookhaven-sized accident. And no insurance company—not even Lloyds of London—

will issue property insurance to individuals against radiation damage. If there is so little risk in atomic power plants, why is insurance so inadequate?

The knowledge that man must henceforth live in constant dread of a major nuclear plant accident is disturbing enough. But we must recognize that even if such calamities are averted, the slow saturation of our environment with radioactive wastes will nevertheless be raising the odds that you or your heirs will fall victim to one of a multitude of afflictions. There is no "threshold" exposure below which we can feel safe.

We have little time to reflect on our alternatives, for the moment must soon come when no reversal will be possible. Dr. L. P. Hatch of Brookhaven National Laboratory vividly made this point when he told the Joint Committee on Atomic Energy: "If we were to go on for 50 years in the atomic power industry, and find that we had reached an impasse, that we had been doing the wrong thing with the wastes and we would like to reconsider the disposal methods, it would be entirely too late, because the problem would exist and nothing could be done to change that fact for the next, say, 600 or a thousand years." To which might be added a sobering thought stated by Dr. David Price of the U.S. Public Health Service: "We all live under the haunting fear that something may corrupt the environment to a point where man joins the dinosaurs as an obsolete form of life. And what makes these thoughts all the more disturbing is the knowledge that our fate could perhaps be sealed twenty or more years before the development of symptoms."

What must be done to avert the perils of the peaceful atom? A number of plans have been put forward for stricter regulation of activities in the nuclear utility field, such as limiting the size of reactors or their proximity to population concentrations or building more safeguards. As sensible as these proposals appear on the surface, they fail to recognize a number of important realities: first, that such arrangements would probably be opposed by utility operators and the government due to their prohibitively high costs. Since our government seems to be committed to making atomic power plants competitive with conventionally fueled plants, and because businesses are in business for profit, it is hardly likely they would buy these answers. Second, the technical problems involved in containment of radioactivity have not been successfully overcome, and there is little likelihood they will be resolved in time to prevent immense and irrevocable harm to our environment. Third, the nature of business enterprise is unfortunately such that *perfect* policing of the atomic power industry is unachievable. As we have seen in the cases of other forms of pollution, the public spirit of men seeking profit from industrial processes does not always rise as high as the welfare of society requires. It is unwise to hope that stricter regulation would do the job.

What, then, is the answer? The only course may be to turn boldly away from atomic energy as a major source of electricity production, abandoning it as this nation has abandoned other costly but unsuccessful technological enterprises.

There is no doubt that, with this nation's demand for electricity doubling every decade, new power sources are urgently needed. Nor is there doubt that our conventional fuel reserves—coal, oil, and natural gas—are rapidly being consumed. Sufficient high-grade fossil fuel reserves exist, however, to carry us to the end of this century; and new techniques for recovering these fuels from secondary sources such as oil shale could extend the time even longer. Furthermore, advances in pollution abatement technology and revolutionary new techniques, now in development, for burning conventional fuels with high efficiency, could carry us well into the next century with the fossil fuels we have. This abundance, and potential abundance, gives us at least several decades to survey possible alternatives to atomic power, select the most promising, and develop them on an appropriate scale as alternatives to nuclear power. Solar energy, tidal power, heat from the earth's core, and even garbage and solid-waste incineration have to some degree been demonstrated as promising means of electricity generation. If we subsidized research and development of those fields as liberally as we have done atomic energy, some of them would undoubtedly prove to be what atomic energy once promised, without its deadly drawbacks.

Aside from the positive prospect of profitability in these new approaches, industry will have another powerful incentive for turning to them; namely, that atomic energy is proving to be quite the opposite of the cheap, everlasting resource envisioned at the outset of the atomic age. The prices of reactors and components and costs of construction and operation have soared in the last few years, greatly damaging nuclear power's position as a competitor with conventional fuels. If insurance premiums and other indirect subsidies are brought into line with realistic estimates of what it takes to make atomic energy both safe and economical, the atom might prove to be the most *expensive* form of energy yet devised—not the cheapest. In addition, because of our wasteful fuel policies, evidence indicates that sources of low-cost uranium will be exhausted before the turn of the century. Fuel-producing breeder reactors, in which the nuclear establishment has invested such high hopes for the creation of vast, new fuel supplies, have proven a distinct technological disappointment. Even if the problems plaguing this effort were overcome in the next ten or twenty years, it may still be too late to recoup the losses of nuclear fuel reserves brought about by prodigious mismanagement.

The proposal to abandon or severely curtail the use of atomic energy is clearly a difficult one to imagine. We have only to realize, however, that by pursuing our current civilian nuclear power program, we are jeopardizing every other industry in the country; in that light, this proposal becomes the only practical alternative. In short, the entire national community stands to benefit from the abandonment of a policy which seems to be leading us toward both environmental and economic disaster.

Man's incomplete understanding of many technological principles and natural forces is not necessarily to his discredit. Indeed, that he has erected empires despite his limited knowledge is to his glory. But that he pits his ignorance and uncertainty, and the fragile yet lethal technology he has woven out of them, against the uncertainties of nature, science, and human behavior—this may well be to his everlasting sorrow.

9

"Bananas in Vermont"

by Richard M. Klein

"Bananas in Vermont" is the first of a two-part article by Richard M. Klein. Dr. Klein, who has followed closely the development of atomic power in Vermont, advocates turning the potential disadvantages of nuclear plants into advantages. Professor of Botany at the University of Vermont, Dr. Klein's future projects include research on the effect of light on cells, a book on plant physiology for gardeners, and a visit to Southeast Asia. With his wife—also a plant biologist—he is the author of Discovering Plants *and* Research Methods in Plant Science.

Vermont is not a wealthy state. Industry produces the largest share of state income, tourism is second, and agriculture third. It requires automation plus 40 to 70 head of cattle to make a go of the family farm, and the number of farms is decreasing each year. In the northeastern part of the state, the "northeast kingdom," where the *true* Vermonter traces his ancestry back at least five generations, some residents live close to the subsistence level; jacking of deer provides table meat. Even in the larger cities underemployment is not uncommon.

Vermont cities are not large by California or New York standards, but some, like Burlington, are seeing the development of middle-class suburbs whose residents, mostly "instant Vermonters," are putting increasing pressures on roads, sanitary facilities, public services, and, overwhelmingly, on schools. These demands on available funds can, lacking an industrial base, be relieved only by pushing up property taxes and, as was recently done, by imposing a greatly resented sales tax.

New sources of income must be found and tapped. The most obvious of these would be to bring in new industry sufficient to employ those who leave the farms and to provide a new tax base; and the second, to use the land more extensively and better exploit its natural beauty by attracting people who can build summer and, eventually, permanent homes. Many of you have seen the ads extolling "Vermont, the Beckoning Country," a publicity venture of the Vermont Development Commission. Having recently moved to Vermont, I would just as lief stop immigration altogether, although I realize that this would result in action by the federal government, an organization with more firepower than we can muster even with the muzzle-loaders now selling for exorbitant prices in our antique shops. Independent of my desire to trout fish twenty minutes from home, it is obvious that both industry and land development will be encouraged.

Despite the church mouse condition of the state treasury, Vermont is a geographic area rich in natural beauty. Its valleys are bucolic remnants of the mid-nineteenth century; the Green Mountains are soft, and, in contrast with the Rockies or the Alps, are "take-in-able." International Business Machines recognized this a few years ago when it decided to open a facility in the Burlington area, thus enabling its employees to find homes in a region unexcelled for those features that were cover paintings in the now-defunct *Saturday Evening Post*. Running parallel to the mountains for almost 130 miles and forming much of our western border is Lake Champlain, where one can still see far down into the clear depths to observe and catch bass, walleyes, perch, and northern pike. There is a deep and abiding love for this environment, and it is doubtful whether most residents would knowingly despoil the land or the water. Yet, good intentions are inadequate safeguards and it is instructive to examine more closely the problems inherent in reaching goals dictated by economic necessity.

For industry to be attracted, it must be assured of an intelligent work force, land, good and abundant water, easily reached markets, and adequate power. In our modern transport-oriented society, sources of raw materials need not be a major problem; pipelines could bring crude oil from deepwater ports to refineries on the shores of Lake Champlain or even on top of Mount Mansfield. This is an extreme, yet there are residents who would be reluctant but complacent boosters for heavy, pollutant-producing industry—"My taxes would drop and it wouldn't be in *my* backyard."

Logically, the industrial ideal would be clean, nonpolluting, skilled fabrications, but this requires power. Vermont, singularly blessed with many attributes, has few potential hydroelectric sites, there is no native coal, and present power consumption in the northeast is so great (remember the November, 1965, blackout?) that the addition of more industrial consumers of power would overwhelm present facilities. Burlington just obtained voter approval for an oil-fired turbine to serve as a backstop for its municipally owned coal-burning generator. Incidentally, the city fathers in all their wisdom will site the new plant in an existing building directly on the lakefront where there are only two beaches for the whole area, both inadequate and one now closed because of sewage pollution.

As Sen. George Aiken has pointed out, the answer to our power bind is atomic energy. Recognizing the profit inherent in serving new industry, a consortium of private power companies has already started construction of one plant on the Connecticut River near the village of Vernon, and this plant, plus contemplated or completed units throughout the country, must serve as a model for a Lake Champlain plant, which may eventually be built at or near the town of Charlotte, just south of Burlington in Addison County.

Assuming, for a moment, that another atomic energy plant is inevitable, it is obvious that it is the task of aroused citizenry and concerned scientists to direct their efforts toward setting realistic limits on the design and operation of such a plant. Articles in *Natural History* and other publications have detailed potential health dangers and environmental insults that accompany the operation of any nuclear plant. Others have shown that the Atomic Energy Commission is, like many other major federal agencies, capable of dissembling when it is useful to do so. As several sympathetic opponents have said, it is most difficult for the same agency to be directed by law both to promote the use of the atom and, at the same time, to regulate its use. Here lies schizophrenia with a touch of defensive paranoia, and the Atomic Energy Commission is exhibiting symptoms of both these mental diseases.

Like the Pentagon, the AEC has scarcely been questioned for close to 20 years, and in this period of time it has consolidated an extensive power base. Yet, in only three years, there has been a series of blows to its hegemony. One hard blow was delivered by a new activist group formed in upper New York State and centered about Cornell University. Briefly, AEC and several power groups decided to construct a fairly large plant on Lake Cayuga. The boiling-water reactor was to be over 800 megawatts in capacity, and like other plants, it was to follow a design established by AEC, with safeguards and performance standards set by AEC, and with supplementary funds provided by AEC. Prof. David D. Comey and his colleagues presented—in public hearings and in two excellent publications—a serious study of the inherent and unpublicized dangers of such a plant, and finally got the AEC to withdraw its approval pending restudy. More recently, the Minnesota legislature insisted that the state can set more stringent and restrictive pollution standards than those of the AEC, a position the commission finds, at best, uncomfortable and, at worst, unbearable.

With, perhaps, a less firm professional base, a group of citizens from Vermont and New York have formed a Lake Champlain Committee dedicated "to eliminate water pollution from all sources . . . and to conserve the natural resources and scenic beauty of the Champlain Valley." The Vermont-New York Committee promoted and obtained passage of landmark legislation in both states to provide that smaller administrative units (like Charlotte) cannot be pressured into an environmental blunder. The temptation for a town can be strong: the tax intake in a town that has an atomic energy plant would vastly increase, while its tax rate could be substantially reduced. Citizens can picture new parks, new swimming pools, new municipal buildings, new schools and, of course, higher property values.

Senator Aiken and Governor Davis arranged for the AEC to set up a public meeting at the University of Vermont. There, many of their big guns spoke on the safety, necessity, and value of atomic energy plants in general and for Vermont in particular. The federal armamentarium included a Nobel laureate, other members of the commission, section chiefs, and a diverse sprinkling of scientific talent from the Oak Ridge National Laboratory. An expensive exhibit was brought up from Washington, miniskirted usherettes with white blazers lettered "AEC" were much in evidence, and the rostrum was graced with the AEC seal as well as those of the state and the university. Vermont, population about 400,000, was to get the "big" treatment.

Morning seminars given by their scientists were standing room only, and as a university teacher I was delighted with the educational experiences that my students got. The presentations were, to be charitable, below the standards that we set for student seminars; expected information was not forthcoming, and many in attendance got the impression that we were boondocks residents to be patronized. The formal afternoon presentations were held in the university gymnasium and attended by the academic community as well as by the general public. Remember, please, that town meetings are part of our way of life. The AEC was clearly running scared and the soothing speeches of commission members and their scientists were closely followed by the audience. After the AEC had two-thirds of the program, a panel of four "environmentalists" joined the commissioners for a session of screened audience questions. It was at this point that AEC's careful plan went awry, for the questions were sharp and to the point (remember the town meeting syndrome), and the sympathy of the audience began to shift away from the official line and toward the Vermont tradition of believing what makes "good sense." This was the first time that the AEC had mounted a massive public meeting, and one of the few times that variant opinion was allowed to share the same platform. Certainly the commissioners were talking to the nation here, not just to Vermont, and certainly they were defensive and a bit apprehensive. Governor Davis received prolonged applause when he concluded the afternoon's session with a firm statement that "the right to increase safeguards should be reserved to the state." Vermont has now submitted a friend-of-the-court brief in Minnesota's suit—a development not planned by the AEC.

In October, Vermont Yankee Nuclear Power Corporation teamed up with the AEC for a two-day public meeting with the citizens of Brattleboro, near the site of the nuclear power plant under construction on the Connecticut River. High school and college students as well as local citizens attempted to get some firm information to allay their fears about this plant but, regrettably, heard only sophistries and "facts" that are questioned by many ecologists. Again, AEC struck out. At about the same time, a committee of the state legislative council met to draft new laws on energy and power. Vermont Yankee officials succeeded in watering down the strong recommendations to the legislature. There was, they said, no quarrel with the need to protect the environment, but power needs had to be paramount.

Nongovernmental scientific opinion suggests that standards of maximum allowable exposure to radiation should be reduced, which would be technically possible with appropriate changes in plant design. More stringent standards will undoubtedly increase the construction cost of a generating plant by dollar amounts that are huge at first glance. What is not pointed out, however, is that when the excess costs are amortized over plant life, the additional expense will add only a small fraction to the cost of each kilowatt hour generated and sold.

Thermal pollution is, however, another matter. Consequences of pouring millions of gallons of heated water into any body of water are still not completely known; but to assume that because we don't know the consequences we can move right in, is ecologically dangerous and scientifically indefensible. A 500 kw. reactor cooled by lake or river water must pump 5,600 gallons of cooling water *per second.* As AEC reluctantly admits, there will be alterations in the biological and physical nature of a relatively slow-moving lake like Champlain and there will undoubtedly be changes in the surrounding countryside as well. Northern Vermont is very cold in the winter, and it is a rare year that Lake Champlain doesn't freeze over. Warmed-up, it may not freeze over completely. As cold air, borne by the prevailing westerlies, comes whistling down from the Adirondacks and crosses a stretch of open water, it will pick up sufficient moisture to produce severe fog and sleet. Addison County, self-styled "land of milk and honey" (which in fact it is) may spend six months as an inland equivalent of the Outer Banks.

Of course, if sufficiently pressed, the designers can include cooling towers or recycle their cooling water as a car radiator does. These changes would cost up to six million dollars, but in terms of total outlay they are not economically impractical. It certainly will help preserve the lake, but it will provide the makings of fog not only during the winter—if the towers can, indeed, work in the 20-below weather—but in the summer as well. And if cooling towers do not handle all the water needed for the plant, do we go back to the lake?

A bit of imagination, apparently lacking in AEC as it is conspicuously lacking in the Corps of Engineers, might just result in better solutions than the standard hot-water-into-the-lake concept, or even cooling towers. One alternative to the cooling tower is the construction of a cooling pond, which would have the advantage of permitting location of the generating plant away from the lake entirely. This would require about 1,500 acres of surface water, and if land prices were reasonable and construction costs low, the added charge would be about $1.50 per kilowatt hour, a cost competitive with cooling towers.

It seems to me, however, that we can do more with pollution problems by making them pay off economically than by wringing our hands or letting massive federal bureaus write off an environment in their pursuit of power. Philadelphia is now packaging its trash, putting it into empty coal cars and then shipping the material back to the strip mines as land fill. The city saves a million and a half dollars a year by avoiding incineration, and the Reading Railroad makes money as well.

Heat from an atomic reactor is available in tremendous supply. If it is used instead of wasted, several related problems can be solved. We need jobs for agriculturally trained but underemployed, people; we need new taxable industry; we need a clean lake; and we need intelligent land use. The Champlain Valley is good farmland, but the growing season is so short that production of truck crops for the big city markets is impractical. Cantaloupes, for instance, produce a crop only occasionally —a risky business, indeed. Suppose that the waste cooling water were piped away from the reactor and used to heat large greenhouses that could produce tomatoes, melons, cucumbers, and other crops in the winter? Thermal heat exchangers could produce air-conditioned areas during the summer to permit mushroom cultivation. Not only could the power companies sell a waste product pollutant at a profit, but they could sell electricity to light the greenhouses. These are not just pipe dreams; Iceland has been using thermal spring water for many years for precisely these purposes, as well as for heating homes in Reykjavik. Stockholm has placed an atomic energy plant within the city limits and uses the cooling water for home and office heating. Rough calculations by a nonengineer (me) have indicated that even a modestly sized plant can produce enough waste heat to make a portion of the Champlain Valley into an artificial subtropical paradise. Still a land of milk and honey—plus tomatoes, mangoes, and jobs. Might we not also have girls in grass skirts undulating beneath banana trees?

10

"Green Mountains, Green Money"

by Richard M. Klein

"Green Mountains, Green Money," a discussion of the problems and pressures of land development in Vermont, concludes Richard M. Klein's two-part article, "Bananas in Vermont." Dr. Klein emphasizes that he is writing as an individual, not as a representative of the university or of the state of Vermont.

The state of Vermont and its people have a very favorable image throughout the United States. Visions of frosty mornings and horses pulling sledges of maple sap to a steamy sugar house, the Long Trail in the summer and ski trails in the winter, hillsides ablaze with autumnal colors, and residents with a reputation for honest taciturnity, frugality, and homespun hospitality make people beam when you say you are from Vermont. Surprisingly, all this is true: Vermont is a *Reader's Digest* ideal. Yet the state is caught in the grip of another part of the American ideal, the concept of rising material expectations, and is wrestling with the question of how to pay for it.

One solution attempted over the past few years has been to attract new industry whose activities will not destroy the land, the water, or the people. This has been reasonably successful; today, industry brings in more money than farming, and there are now more people in the state than cows—a reversal of the situation that prevailed not too long ago. To continue to attract clean, nonpolluting industry, electric power is required, and the current furor over the construction of atomic energy generating plants in Vermont is an ecological example of the academic "on the one hand, but then again, on the other. . . ." It is likely that there will be a second atomic energy plant built in the near future, but it is still an open question whether it will be rammed down our throats by Vermont Yankee Power and the Atomic Energy Commission or be designed so as to truly limit inherent environmental insults and provide additional benefits to the land and the people.

Utilization of land in Vermont presents about as many problems as does power. A short growing season, shallow soil, traditional rockiness of the fields, and abrupt transitions in elevation all dictate that much of the state is agriculturally marginal. Large tracts in and near the Green Mountains were held in timber, culled to sugar maple, or left alone with occasional selective cutting of timber. Up to about the end of World War II, land simply wasn't worth much; it went for less than $20 per acre. Land deeds are so written that acreage is casually given as so many acres "more or less" and a 10 percent variation, usually greater than smaller, is expected. Because of depressed valuation, land assessments and land taxes brought in relatively little money. Even today, Vermont's Department of Forests and Parks owns over 100,000 acres in the state and in 1968 paid town taxes of about $.31 per acre. In one town, Groton, the state's holdings constitute more than 20 percent of the tax listings. These are fine recreational lands; if they were taxed according to their fair market value, Groton would be wealthy—and the town knows it.

Four lumber companies—St. Regis, Brown, Atlas Plywood, and International Paper—own close to 800,000 acres of land in Vermont. Under sustained-yield forest management, profit on an acre of timberland is about $1.50 per year. When land is assessed at its fair market value, the taxes on such land can easily be higher than the timber income. Landowners and towns realized that there is both immediate and long-term money in developing the land for recreation. With this realization, some towns are raising land taxes to force owners to sell for development, and many towns are under pressure to sell off their municipal forests for both immediate cash and long-term tax gains. The land boom is definitely on.

Land development and land speculation are old hat in the United States, and Vermont is no exception. The state's founding fathers, Ethan and Ira Allen, pushed their Onion River Land Company between bouts with the British. The first governor of Vermont, Thomas Chittenden, speculated in real estate while in office. Small-scale developers came and went for a hundred years or so—their avarice or their honesty had small effect on the over-all stability of the environment.

Today's activities are, however, on a scale several orders of magnitude greater than anything ever seen before. The affluence of at least part of the population, the decreased livability of the eastern seaboard's megalopolis, the extension of super-highways and the cars that use them, and the desire of many people to breathe fresh air at least part of the year—all these factors mesh neatly with the situation in Vermont where agricultural and forested lands are economically unprofitable, and where the towns and the state are desperately searching for a firmer, broader tax base to meet the spiraling desire for government services. It is no longer a question of whether Vermont land will serve as the environmental safety valve for city people and as the economic salvation of the towns, but rather of who will provide the developmental services and how will such development be controlled.

Vermont is divided into fourteen counties. Windham County, my focus of attention, includes the southeastern part of the state with the Connecticut River forming its eastern margin. Brattleboro, population 13,000, is the only community of any size, with small towns, villages, and hamlets dotting the rest of the county. The county is very lovely. It boasts eight ski areas, including the renowned Mount Snow-Carinthia-Haystack complex and Stratton Mountain. Thousands of well-heeled visitors (a pair of skis can cost $200; boots, $150) receive their first taste of Vermont through skiing visits. Green Mountain National Forest, several reservoirs, excellent trout streams, and a series of tranquilizing vistas have stimulated the desire of many people to own a piece of Windham County.

Local officials thought that they were knowledgeable about summer people. In 1968 there were 3,246 vacation homes (called "camps" by real Vermonters) in the county; Vermonters owned less than 20 percent of these. People from Connecticut, Massachusetts, and New York owned the rest. About $600,000 was realized from property taxes; the average tax bill per camp was less than $200, and none of the towns got rich on that. Of course, purchases and services brought in additional money. At the end of 1968, there was greater appreciation of summer residents since they accounted for one-quarter of all revenue. As of the fall of 1969, the situation was very different; it can best be evaluated by focusing on two townships in Windham County, Dover and Stratton. The situation in these towns is atypical only because things have moved faster here than elsewhere in the state.

The town of Dover is just a few miles from the Mount Snow-Carinthia-Haystack Mountain ski complex. There are 23,000 acres in Dover Township with 2,000 acres tied up as national forest. Four hundred people are registered voters. Up to about 1955, Dover's economy was agricultural with considerable lumbering, but today only one farm is operative and lumbering is minimal. In keeping with its location, its facilities, and its economic position, the annual town expenditures and the decisions involved in disbursing its funds were not onerous for the town clerk and three unpaid selectmen. The spectacular success of the ski areas resulted in additional revenue; ski patrons and ski bums were in, but not of, the town during the winter and generally disappeared with the snow. There were 300 summer residents as of 1968, and the possibility of an increased number of camp owners in colonies surrounding the ski areas was viewed with interest. The Mount Snow Development Corporation owned and planned to develop several moderately sized areas, and other ski companies were also considering this prospect to provide year-round use of their facilities.

With these possibilities in mind, the town had published a zoning ordinance in January, 1967, to ensure the orderly development of summer homes. The minimum requirements for lot and house size, setback from roads, access, and so on were in keeping with tradition and common sense, but they were just that—minimal. Except for the general statement that there should be no hazards to health and that the creation of unsanitary conditions was to be prevented, health and safety regulations were essentially lacking. Basically, Dover does not have the manpower or the expertise to regulate home development, and the town finessed with the hope that state standards and controls would fill in the gaps.

Including the ski areas, there are now over twenty developers in Dover holding close to 7,000 acres. In 1966, West Virginia Pulp and Paper Company sold 3,400 acres of timberland to the Pineland Realty Company, and this parcel was subsequently sold to the Vermont Lumber Company. Eugene Coleman, an active land buyer, acquired Vermont Lumber shortly thereafter and initiated the development of Dover Hills. He sold out quickly to the Cavanagh Leasing Company of New York and Florida. Cavanagh Leasing and Caveland Equities are both directed by Mr. Coleman; stock, which sold for $9 a share when first issued at the end of 1966, was worth $70 per share in August, 1969.

Vermont Lumber essentially went out of the lumber business in 1968 when they announced that they were going to develop 550 acres of their total acreage. If we eliminate land already in homes and discount roads and land that is too steep or otherwise unsuitable, Vermont Lumber owns one-sixth of the land area and controls about one-half of the future of Dover. Decisions affecting the future of the town are out of the hands of the people themselves.

In the spring of 1969 an attempt by the townspeople to restrict building height, control road frontage, and regulate building permits was countered when urban developers swamped the town meeting with company lawyers, a company planning consultant, and $600 in slick publications. The tidal wave of talk by this outside talent effectively packed the meeting, and the local residents—who are the only voters—didn't have a chance. One townsman, Jack Veller, has said that those who will determine the nature of the environment have no sense of Vermont history, no love of land, and no permanent interest because they will never live in Dover.

Of Vermont Lumber's six planned developments, Dover Hills and Dover Hills West are the biggest and are being pushed the hardest. Although lots of one to twelve acres are for sale, the average lot size is just over one acre and the average price is between $5,000 and $6,000. Even at these prices which, incidentally, have greatly inflated the base for tax assessments at "fair market value," about three-quarters of the lots were sold by August, 1969. Assuming that just Vermont Lumber sells all of its acreage, a town of 400 voters will include 3,000 additional homes. Camp owners cannot vote, but their impact on the town and their just demands for road clearing in winter, police and fire protection, and administrative services cannot fail to create severe strains in this community.

Dover Hills is being marketed by the hard sell. My letter from their marketing director, Paul Thibert, was accompanied by a slick brochure with pictures of beautiful people lounging before roaring fireplaces, golfers, pretty girls on horseback, trout streams, and uncrowded ski-lift lines (probably taken on a Monday morning). Dover Hills is "close to convenient shopping, schools, houses of worship, theaters, medical facilities." It depends on what you mean by convenient; furthermore, the public schools of Dover just meet the state standards for room volume per pupil, and hospitals are 25 miles away in Bennington or Brattleboro. Vermont Lumber is using telephone contacts straight from the Connecticut telephone directories; they have not had to extend their area of sales pitch. If a prospect is at all interested, the company sends a "sit-salesman" to the prospect's home to sit down with the family, present a slide show, and discuss the matter. If still interested, the client (no longer a prospect) is invited up for an all-expense weekend where considerable pressure is employed to get a 10 percent refundable down payment. Questions about who wears what pants in which family, appeals to sympathy for the salesman's children, discussions of the value of land as a hedge against inflation, and downright rudeness limit the percentage of returned down payments.

The company has gone to some expense to give the client a good show for his weekend visit. New roads run pleasantly through the development. They have cleared the brush and groomed the woods for about one hundred feet back from the roads, most effective for giving the impression of a mature forest. Sales pitches are replete with promises of things to come, including a 70-acre lake, a golf course, stables, a village shopping center, and (affluent society, indeed!) an airport. As of this writing, none of these extras has materialized. Their promotion speaks of "exclusive Ellis Park," but this is a one-acre site with a picnic table and a children's section containing slides and swings like those in many backyards. Because of a weird loophole in Vermont law, salesmen don't have to be licensed by the state, or even be residents, and their promises beyond the contract are as reliable as snow in June. In an effort to allay the town's concern over schools, one of Vermont Lumber's officials indicated that they would consider donating land, but would assume no responsibility beyond this. Fortunately, there has been little cutting in Dover Hills, and a restrictive covenant under consideration may prohibit extensive tree removal, as well as set broad limits of house size, color, and area. One model home, priced at about $23,000, is completed. It is carpeted, but its imitation wood paneling is apparently nailed to bare studs, for the walls give with gentle finger pressure. Few of the owners have erected homes, but those that have been built are unimaginative, out of keeping with the land, or downright ugly. They are mostly precut or modular units ranging in price from approximately $12,000 up to about $45,000 for the more ornate A frames.

The major entrance road into both Dover Hills and Dover Hills West comes off the main road in the village of Dover Hills West. It has a fairly steep grade and it might be exciting to see some 8,000 people trying to get in or out on a snowy weekend; snowbound vehicles have already overtaxed road maintainence crews. The local fire department is composed of volunteers whose equipment is adequate for present needs, but the new homes and open woods present a fire hazard (no hydrants, for example) that is likely to be reflected in insurance rates. Vermont Lumber has apparently given no thought to rubbish disposal, even though new state laws drastically limit expansion of town dumps.

Although population pressures, taxes, aesthetics, and other problems exist in Dover Hills, the basic worries are water and sewerage. Vermont gets 35 to 40 inches of rain per year, and a good share of this percolates into the ground. With Dover Hills' 550 houses, about 3,000 square feet of land per housing unit will be covered by houses and roads, effectively sealing off about 1,650,000 square feet of land surface. Thus, over 3,000 gallons of water per housing unit for each inch of rain will have to be carried off instead of entering the ground water supply: erosion, silting, and minor flooding can result. Dover Hills will not have a consolidated water supply; each house must have its own well. This may be fine for the man at the bottom of the hill, but the man on the top may find that his well will run dry. The valley householder may not be too well off either, because runoff from the top can flood his property.

In a report to the governor's Commission on Environmental Control, my colleagues Hub Vogelmann, Jim Marvin, and Max McCormack discussed the ecology of upper-elevation land in Vermont, defined as areas above 2,500 feet. They noted that these lands are the primary sources of abundant, clean water in the state because forested land holds rain and collects fog moisture. The soils are generally shallow and rocky, and road construction—or even extensive foot traffic—severely alters infiltration and percolation of water. It is enough to note that a fair proportion of Dover Hills is located at about 2,500 feet; indeed, lots with breathtaking views, which command the highest prices, are at 3,000 feet.

Dover Hills will not have a consolidated sewage system; each house will have its own septic tank. Assuming that each person in each house uses 50 gallons of water per day and that there will be five people in a house, the septic tank must handle 250 gallons of effluent daily. You can multiply this by 550 (the number of homes) to get an idea of the volume of waste to be disposed of.

Bruce Watson, the state soils man, has surveyed in the Dover area. The mountain soils are thin, allowing relatively little soil volume and drainage for a leach field. Raw or partly digested sewage from such a field, plus seepage from the septic tanks, may eventually get into the ground-water and contaminate the wells, particularly those on the lower slopes. Septic tanks used sporadically in vacation homes function more poorly and break down sooner than those servicing year-round homes because they become less efficient. Since bacterial action is depressed by cold weather in areas where soils are shallow, overflow of undigested sewage can be expected. A "sniff test" in several developments indicates that health hazards are already accompanied by aesthetic insults. A Public Health Service team has reported that about one-third of all water systems examined in Vermont were contaminated with sewage. Indeed, well pollution in southern Vermont developments has already been reported by distraught vacation homeowners. Undigested effluents can kill trees, compounding the damage.

Before detailing the reactions of the state to such a development as Dover Hills, a look a few miles up the road might be interesting. In neighboring Stratton township there is a major ski resort that may soon have clusters of vacation homes around the slopes. In 1968 Stratton had only 94 vacation homes paying a bit over $6,000 in taxes. The township has well over 23,000 acres of timbered land ripe for conversion into vacation homes, and virtually all of it is owned by the International Paper Company; International Paper owns 60 percent of the town of Stratton.

The company has been under fire from both Vermont and New York because of a paper mill at Ticonderoga, New York, at the lower end of Lake Champlain. This mill has for many years made a cesspool of the adjacent segment of the lake. A new plant may reduce insults to the eye and nose, not to mention the lake, but pending its completion, sulfite sludge keeps pouring into the water. International Paper's local reputation as a conservation-minded organization is minimal.

IPC has excellent corporate administrators. They recognized that their holdings in southern Vermont were Green Mountain "gold," not because of spruce for pulp or yellow birch for cabinet veneers, but for land development. They realized that IPC had no knowledge of land development and that it would be useful to acquire a company with such experience. Thus, in 1968 IPC put out $12 million to acquire American Central Company, a Michigan-based firm that had developed about seventy tracts in nine states. According to *Fortune,* IPC incidentally generated a tax write-off by transferring land to their wholly owned subsidiary. The president of IPC, Edward Hinman, said, "American Central has had an impressive record in its ten years of operation for selecting, planning and developing recreational areas of this type." To round out their entrance into land development, IPC purchased Spacemakers, Inc., a Massachusetts firm specializing in the construction of leisure homes ("eight weeks from start to completion") and the interdigitation of IPC land, IPC forest products, and IPC land development was complete.

No one is sure how IPC's development plans for Stratton leaked out, and it is possible that the light of understanding dawned only when bulldozers began clearing the land. Residents became alarmed in late June about the instant community and transmitted their fears to state senator Edward Janeway and Governor Davis. Davis immediately telegraphed Hinman, IPC president, for a meeting. A local officer of the company first presented the over-all plan to a state commission in late June, 1969. Asked what would be done about water and sewerage, the IPC official replied that each lot owner would drill his own well and install his own septic tank. Asked by a representative of the Vermont Water Resources Department about tests for septic tank feasibility, the official said that no tests had been made. Asked about open space, the official replied that some land would be left undeveloped, but admitted that some of this is "rather on the steep side." Asked about building restrictions, he replied that none were contemplated. Asked what responsibility his company would assume for sewage treatment plants and a community water supply, he replied that a council of citizens might be formed to study community problems. Asked if IPC or American Central would ante up a bond to pay for emergencies brought about by the development, he replied that he couldn't speak for the company. At the end of this meeting someone passed around a brochure of an American Central development in Wisconsin that showed a lake completely surrounded by 60- by 150-foot building lots. Senator Janeway later admitted that he shuddered.

Shortly after this interesting session, the Stratton Planning Commission met with a citizens' group, the Vermont Natural Resources Council, to point out that tiny Stratton was unequal to the task of controlling the mammoth IPC combine. One commissioner said that the company was able to hire all the lawyers around and had even invited local selectmen to go on the company's payroll as consultants. One local conservationist reported that he was warned to get out of the way or he'd be run over when he tried to take pictures of operating bulldozers, but he hired a plane and took his photographs from the air. According to people who attended the meeting, the pictures show a hunk of land cut by roads from bottom to top, with virtually no thought of water flow or other ecological considerations.

Hinman and Governor Davis met on July 12 in a closed-door session with the Environmental Control Commission, but by the end of the session the state was no closer to learning any details of the plans than it had been two weeks earlier. Hinman did say, in a press release, that his company had only the best interests of Vermont at heart and that American Central would work closely with local and state officials during the development of its 23,000 acres. On the basis of this meeting, the governor said the relationship was starting off as one of the finest cooperative ventures ever undertaken with a developer.

Over the weekend, however, a completely new view of American Central came to light. Elbert Moulton, special assistant to the governor, made an incognito visit to Tamworth, New Hampshire, where American Central is developing a ski and beach club on a small lake. Moulton was greeted at the entrance by a full-dressed clown, complete with balloons and streamers, who waved and beckoned to drivers. Posing as a prospect, Moulton got the hard sell with a "weekend-after-the-Fourth-of-July special," promises of no lower limit on cost of home building, favorable financing, and all the rest. The company advertised in Boston newspapers that there was a lake with a natural sand beach, but the saleswoman admitted that they had been hauling in sand for the past week. Moulton, no slouch as a salesman himself—he worked out the

"Vermont, the Beckoning Country" campaign—admitted that he beat a hasty retreat. In a report to the governor, he concluded that if American Central were to develop in Stratton as it has been developing in New Hampshire, Vermont should not allow it. At the press conference the Friday before, neither the president of IPC nor the president of American Central admitted that they even knew that American Central was developing in New Hampshire.

Within two days, things really began to pop. On Wednesday, Governor Davis announced that development of Stratton by American Central had ceased pending revision of its plans, that the IPC vice president for public relations would meet forthwith with the governor to establish closer communications, that aides of the governor would visit other American Central projects within a week, and that the president of IPC had expressed concern over the type of development being done by his firm in New Hampshire. Local newspapers printed unattributed stories saying that International Paper was not aware of the building habits of American Central prior to its infolding into the corporate bosom, and that IPC now realized that it had misjudged public reaction to its Stratton development and had erred both in its public relations and in its reluctance to discuss things frankly with the Environmental Control Commission the previous week. Although IPC admitted that the New Hampshire operation is tawdry, they stated that it is not the best example of American Central's work. Unfortunately, reports are already filtering in that American Central developments on Cape Cod and elsewhere are at the same level.

On July 25, Stratton residents finally learned officially that much of the township would soon become a community of one-acre lots—10 percent down and seven years to pay. The meeting got off to a bad start because American Central didn't show up and a messenger had to be sent up to the mountain to bring a representative down to the meeting. When the townspeople asked for some assurance that American Central would do right, they were told that the development would be a "welcome addition to your town," and that "our company's past record is spotless." It didn't allay their fears to be told that any jump in school-age population would be the concern of the local school board, that the township would double or triple its population in a few years, that 12 miles of roadway are being built for the town to take care of, or that expansion is virtually certain within the next three to five years.

At about the same time, Governor Davis asked a professional planning consultant to meet with American Central to insure that Stratton would be a quality development. By the end of the week, all work was halted pending extensive review and evaluation of the entire operation by state advisers, and by the end of the month, it was announced that plans for the development would be shelved until at least the spring of 1970. The governor hinted that IPC may decide that the development might simply not be worthwhile because of the unfavorable publicity and the adverse reactions of local citizens.

The purchase of large tracts during 1969 has not been restricted to southern Vermont. Interstate highways will soon complete the network between Massachusetts, Montpelier, the state capital, and the Canadian border. It is a short drive through lovely country into the whole northern part of the state, including the hitherto neglected "northeast kingdom." In one of the kingdom's counties, Essex, 90 percent of all land is owned by lumber companies. Caveland Equities, Inc., also directed by Eugene Coleman (who started Dover Hills), purchased close to 3,000-acres in the kingdom for about $100 per acre—a dirt cheap price. Mr. Coleman, a busy man who keeps his money working hard, also purchased 800 acres on yet-unspoiled Lake Seymour, and 2,700 acres in Greensboro. Some undisclosed group apparently took the trouble to look at a recent state highway map (free from the Vermont Development Commission) and purchased 1,600 acres near Interstate 89 in the town of Sharon. A parcel of about 30,000 acres was sold by a financially depressed lumber company to Laird Properties and Laird's New England Land syndicate added about 40,000 acres within a few weeks. Names like the "Great Northern Land Corporation," "Triton Investments" (a California combine), and several ". . . and Associates" are beginning to crop up on the list of purchasers of large acreages.

An interesting sociopolitical note to the burgeoning land speculation in the northeast part of the state is the separatist movement in Quebec. Many of the Anglo-Saxons (anyone who is not French) have money, and this is part of the reason for the agitation. They are beginning to hedge their finances by heavy buying in those areas of Vermont and New Hampshire within 75 miles of the border and are paying not only inflated land prices but also an unfavorable exchange rate of 7 percent.

Many townships have no zoning regulations and, of those with zoning, many have either interim regulations or, as in the case of Dover and Stratton, minimal zoning. Mere zoning doesn't insure sewerage systems, adequate road construction and maintenance, school sites, a municipal water supply, or even conservation of open space. These are the functions of subdivision regulations, and very few towns have even started arguing about them. Ultimately, direction, power, and naysaying rests at the state level because the towns haven't the money, men, experience, or hometown power to cope with the situation. One official of a prominent development company noted that Vermont is a Mecca for his and other companies just because of this.

The realization that something had to be done immediately caused some soul and law searching in state offices. The secretary of state revealed that a 1917 law gave him the power (hitherto unused) to refuse to license any foreign corporation if the refusal would "promote the general good" of the state. Foreign corporation is defined as a firm whose principal place of business is outside Vermont. It was also suggested that municipalities could refuse to take over the roads put in by developers. This would certainly hurt sales because a potential buyer would understandably be loath to purchase property knowing the roads would not be plowed and maintained in the winter. A simple amendment to existing law would require only that the street or transportation portion of the master plan require approval before any work can start.

Up to 1970, Vermont really didn't have an integrated system for regulation, not because previous administrations were unaware of the dangers, but because there was real citizen resistance to imposition of state control. This past summer several local boards informed the state in no uncertain terms to keep its nose out of their communities—they could handle their own problems, thank you very much.

Archaic practices worked against effective control. The tax department was not permitted to inform other state agencies when a large tract of potential development land changed hands. Fish and Game, Forests and Parks, Water Resources, Health, Planning and Development were separate and unrelated. Central Planning could act only when requested by Development. Highways, responsible for many winter road services, could assist a town only when invited. The list could be extended for another half column.

Vermont does have laws regarding pollution, but like those in many other states, they are ambiguous, full of loopholes, and subject to "interpretation." It hasn't helped that two involved departments, Health and Water Resources, were independent and not especially cooperative. A key section of the Water Pollution Control Act states that the Water Control Board may take action against any person who permits wastes to enter water so as to reduce the water below the classification set for it. Another law provides that any person who diverts water (such as drilling a well) and either corrupts it or renders it impure is liable to the municipality. At a hearing last year the water resources commissioner stated that he didn't believe (but didn't know) that the board could press a case against a mass developer. When asked what would happen if he did seek an injunction, he stated that he thought (but didn't know) that the law allowed him to act only after pollution has set in. He was challenged to determine the legal extent of his power, but he didn't try. The commissioner knew that he could assist in design and had to approve town plans for both sewerage and water, but he didn't believe (but didn't know) he could regulate a development's plans unless they were submitted by the town.

The health commissioner admitted that Vermont has many laws that provide for regulation, but he noted that he doesn't have the staff or the money for environmental control. Some time back the commission won an injunction forbidding construction of additional homes on a site that couldn't support more septic tanks, but it hasn't taken advantage of this precedent. Health officers said they really don't get complaints until after damage has been done, but I can show you water and health hazards a half hour from my home that are older than I am. Obviously, money is a factor, but the prime considerations are motivation and some sense of urgency, and both have been conspicuously lacking. Up to now the people haven't leaned hard on the state because they haven't been aware of the situation.

With the citizens now aroused, these deficiencies are being cleared up by an extensive reorganization of state agencies. Concern for the quality of the environment, shouted for years by the conservation-minded, is now as Vermontish as white rat-trap cheese or maple sugar on snow with dill pickles. We now have an Interagency Council on Natural Resources embracing Forest and Parks, Fish and Game, and several others. In September, a Planning and Community Services Commission was organized to umbrella housing, poverty programs, local affairs, the Central Planning Agency and several others. Not incidentally, it is headed up by Ted Riehle, who fought for and eventually secured the passage of an antibillboard law, only the second in the United States, which will allow all of us to see our mountains without eyeball pollution.

The capstone in any concerted program is an environmental control commission; Vermont established one last summer. Chaired by a legislator with a record of intelligent conservation policy, the commission was charged with making recommendations on new laws, on integration among departments, and on coordination with local governments. Among its first recommendations was one for a 90 percent increase in funds allotted to regional planning groups on the county level, a fiscally, politically, and environmentally sound idea. The commission serves now as an evaluative center for other far-reaching plans. The health commissioner requested the Environment Control Commission to seek legislation that would require all land developers to submit plans to his department for approval prior to any work. He also asked for legislation to give the health department specific authority to move against developers.

Should the state set standards for approving or disapproving certain lands for development? Should Health and Water Resources hire engineers to help towns control land development? Can taxes be scaled to provide advantages if land is not used for development, and what effect will this have on state and local financing? What about pesticide controls, responsibility of town and state for highway construction, protection for lands above 2,500 feet? An important and delicate question is how much authority the state can assume without infringing on local communities. Many argue that local officials cannot be trusted with authority over land development because they have been put on the payrolls of developers, may be in a position to profit personally from development, or are snowed by the high-priced talent at the command of the developer.

In late June, the governor announced that the Environmental Control Commission will form Development Technical Advisory Teams to provide assistance and advice to communities and planning commissions. Happily, the commission was flooded with volunteer expertise, including nationally known regional planners, a professor of zoning laws, and others who are either "instant Vermonters" or "sunshiners" (owners of vacation homes). Economists, ecologists, and other academic types are also available to the communities—not that they weren't before, but few ever asked for their help.

Apparently, the response has been excellent. The town of Wilmington, just south of Dover, asked for a Technical Advisory Team within a week or so after the teams had been set up. A few days later the Stratton Planning Commission sent in an SOS. There are indications that at least some of the developers will avail themselves of these services. In fact, the promulgation of new health-safety regulations that require testing for percolation, groundwater levels, and other basic data will provide, for the first time, a basis for the teams to work effectively. A proposed revision of the sanitary engineering regulations, subchapter on subdivisions, provides the teeth to bite off the outstretched fingers of greedy developers.

I am usually unsatisfied by upbeat endings. Lovers wending their way into the sunset to swelling chords leave me cold. And yet, a colleague who has been deeply involved in Vermont conservation recently told me that he has never been so optimistic about the cause, and I respect his judgment; he has more battle scars than a Roman gladiator. If there is light at the end of this long environmental tunnel, it is the result of the long-overdue realization by Vermont citizens that the "beckoning country" must exist for the future. In a lecture about two years ago, William O. Douglas, a mean infighter for conservation, said that only the people can save their environment, and only if they will support those who have long cried in the wilderness, and only if their wishes are made clear to their elected representatives. These case histories in Vermont tend to support his contention. But public opinion and public support are fickle things. . . .

11

"Dwindling Lakes"

by Arthur D. Hasler and Bruce Ingersoll

Arthur D. Hasler is an ecologist and an authority on eutrophication, the process of lake aging. Author of more than 100 articles on limnology, fishing science, and animal behavior, he also originated the theory that salmon return to their home stream to spawn by identifying a dilute odor characteristic of that stream only. Dr. Hasler is now Professor of Zoology and Director of the Laboratory of Limnology at the University of Wisconsin. In addition to fishing, he finds diversion playing the French horn in the Madison Symphony Orchestra.

Co-author with Dr. Hasler of "Dwindling Lakes," Bruce Ingersoll has been a lumberjack in the forests of Washington's Olympic Peninsula, a guide for fishermen in Wyoming and Minnesota, and a civil rights and crime reporter for the Chicago Tribune. He received his M.S. in conservation communications at the University of Wisconsin this year, and is currently an environment writer for the Chicago Sun-Times.

Algae, minute and one-celled, ride the summer waves of every waterway in America. They colonize stagnant sloughs of rivers and gather in backwaters behind dams. Under a microscope, these tiny and rather attractive plants appear innocent enough, yet they can quickly cover a bay with scum, form hairy filaments to enslime a rocky shoreline, or clog an entire lake. With life-spans of only a few days, algae can make any lake grow old thousands of years before its time.

To realize their immense potential for harm, however, algae depend on man. It takes man to speed up natural eutrophication, the normal process of enrichment and aging undergone by bodies of fresh water. By fertilizing the nation's waters with the nutrients vital to algae growth and reproduction, primarily nitrogen and phosphorous, we turn eutrophication into an accelerated, cultural process—cultural in that we are perverting nature with municipal sewage, industrial wastes, agricultural drainage, and other odious by-products of our civilization. Cultural eutrophication, therefore, is an aberration: a natural process running amok.

In enriching the water with the nutrients in sewage, groundwater, and urban and rural runoff, we promote the rapid reproduction of algae: the seemingly harmless alga becomes 2 algae, then 4, 8, 16, 32—multiply-until there are billions. And during such a population explosion, lakes become murky and fetid under the August sun, while wave-tossed weeds, bloated fish, and dead algae rot in shoreline windrows.

More than one-third of America's 100,000 lakes are showing signs of cultural eutrophication. The danger of accelerated eutrophication continues to grow as our population makes greater demands on national water resources, and lakes continue to take on the euthrophication syndrome. Besides showing the symptoms of excess nutrients and algae bloom, they are characterized by the depletion of dissolved oxygen in deeper waters, a change from cold-water game fish to "rough" bottom feeders, and the encroachment of rooted vegetation from shore.

The appearance of these symptoms follows a definite sequence. First, the algae population skyrockets in a man-fertilized lake. Water fleas (miniature freshwater shrimp), the staple in the diet of fry and minnows, cannot eat enough algae to keep these plants in check. As a result, billions of algae live their languid lives, reproduce, and then die. As they drift toward the bottom, their decomposing bodies exhaust the deep-water oxygen supply. Trout, whitefish, and other fish species suffocate in the oxygen-thin depths.

The lake's ecology, initially upset by excess nutrients, then becomes totally upended, since bacteria can convert only some of the dead algae into plant and animal food. Therefore, generation after generation of algae settle on the bottom, adding layer after layer to the muck. The rate of sedimentation is most rapid in a northern lake where bacteria grow only during the summer while nutrients are added throughout the year. As erosion and sedimentation fill in the lake, shoreline vegetation impinges on the open water. In time, the lake becomes so shallow and overgrown that it becomes a marsh or bog. The accelerated process of aging has taken its toll: the lake's life is ended.

Cultural eutrophication is not new, nor is it solely an American problem. Recent core samples of an Italian lake indicate that ancient Roman road builders caused eutrophication by exposing nearby nutrient-rich limestone strata to erosion.

By the latter part of the nineteenth century, a few scientists in Europe and the United States, alarmed about cultural eutrophication, issued warnings that went unheeded. In 1896, on Switzerland's Lake

Zürich, the problem finally received notice, and during the 1920's and 30's worldwide scientific interest began to focus on the problem. But research without implementation is impotent, and by the 1940's eutrophication was no longer confined primarily to farm belts and urban areas. It had followed urban man in his quest for recreation into the wilds.

It took the visual (and olfactory) impact of a huge body of water, Lake Erie, suffocating as a sump for industrial waste, sewage, and urban and rural runoff to bring the problems of water pollution dramatically to the public eye. Some now pronounce Lake Erie "dead." We, however, extend hope for recovery. For if communities and industries in Lake Erie's drainage basin cease polluting and fertilizing its waters, and if a cleansing flow from Lake Huron is permitted to reach and "flush out" the lake via the Detroit River, Erie might show signs of recovery within a decade. True, our hope is contingent upon many "ifs." However, it cannot be too strongly emphasized that these situations are reversible—that bodies of water will respond when the right steps are taken.

There is reason to fear for the other Great Lakes, together the greatest reservoir of clean fresh water in the world. Although Lake Superior and Lake Huron bear only a few signs of eutrophication so far, alarming amounts of DDT and other persistent toxic pesticides have been detected in the highly prized flesh of lake trout caught in their waters.

Lake Michigan is also deteriorating rapidly from increasing effluents and may soon go the way of Erie. The strench of algae and weeds decaying on beaches, compounded by the wave-borne plague of dead ale-wives, makes eutrophication impossible to ignore in this once pure lake. Such cities as Muskegon, Gary, Chicago, Milwaukee, and Green Bay realize that the fate of Lake Michigan—to which they once owed their existence—is now uncertain.

Stewart L. Udall said a year ago that there is still time to save Lake Michigan, but warned that further delay in action would prove fatal. His statement is admittedly, and of necessity, vague, but it is correct.

The long-range outlook here is not encouraging because the mitten-shaped lake isn't in the mainstream of Great Lakes water circulation, and thus receives very little cleansing flow. Its tributaries carry heavy nutrient loads to its shallows, and consequently, along Michigan's southern rim scientists have found 44 times more phosphorus (in the form of sewage effluent and industrial waste) than the lake can handle.

Lake Michigan's fate would have been decided long ago had Chicago's engineers and city fathers lacked the foresight and imagination to reverse the lakeward flow of the Chicago River and channel it into

STAGES OF NATURAL LAKE SUCCESSION

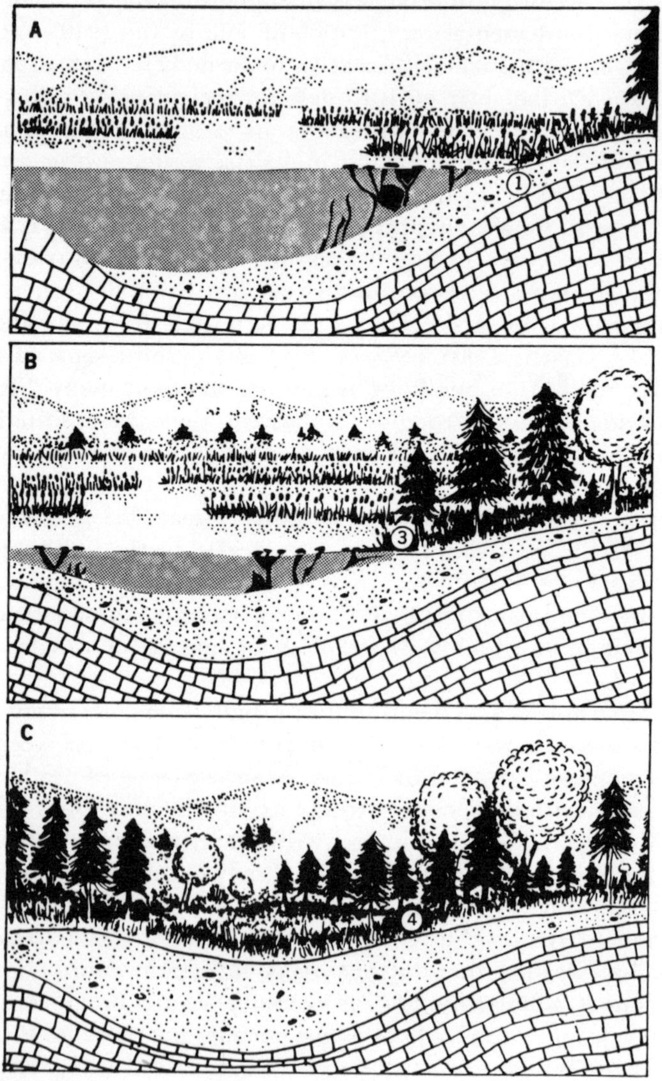

FIGURE 11-1 (A-C): portray the natural aging and disappearance of a lake in the Northern Hemisphere. A newly formed lake contains few algae or eroded materials. Later, shoreline runoff brings eroded soil (1) and plant nutrients into the lake; aquatic vegetation (2) (including algae) flourish and die, adding to debris on lake bottom; marsh plants and water-tolerant conifers (3) grow at lakeside. Eventually, the entire lake is filled and covered with forest vegetation (4).

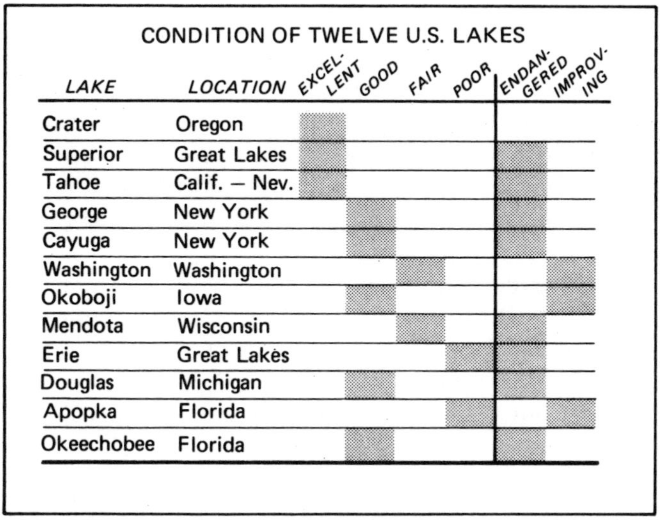

FIGURE 11-2: Condition of twelve U.S. lakes.

the Illinois River, a tributary of the Mississippi. The metropolis now draws water from the lake, and discharges sewage effluent into the altered river system. As is sometimes the case in such diversion operations, they have alleviated their own problem but have contributed to the water problem of St. Louis and other cities located farther down the Mississippi.

While concern for the future of the Great Lakes intensifies, distress is being voiced over eutrophication on many smaller bodies of water. Fishermen complain that dense mats of algae and rooted weeds make trolling for pike impossible, and that the few fish they do manage to catch are too tainted to eat. A summer cottager realizes that he will be landlocked if his bay becomes any shallower, and a concerned parent forbids her youngsters to wade in the green slurry of algae.

To stop cultural eutrophication, the sources of nutrients for algae growth and reproduction must be pinpointed. A 1967 survey, representative of the Midwest, indicates that 6 per cent of the nitrogen and 2 per cent of the phosphorus reaching Wisconsin waters come from septic tank seepage. Twenty-five per cent of the nitrogen and 56 per cent of the phosphorus come from municipal sewage treatment plants. Runoff from manured fields accounts for 10 and 22 per cent of these elements, while urban runoff supplies 6 and 10 per cent. Groundwater and direct rainfall on Wisconsin bodies of water contribute, respectively, 42 per cent and 9 per cent of the nitrogen, but together less than 4 per cent of the phosphorus. Industrial wastes which bypass municipal treatment are a source of trace amounts of each of these chemical substances.

The nitrogen contribution of groundwater and rainfall will continue to be significant as long as automobile exhaust and industrial smokestacks keep spewing nitrogen into the atmosphere. Rain, in cleansing the air, picks up this nitrogen and deposits it directly into lakes and streams or indirectly in groundwater. Nitrogen-laden rain falling on land also percolates down to the subterranean water table, the level of groundwater.

It must be remembered that the natural rate of eutrophication is the base rate for cultural eutrophication. Moreover, the natural rate depends on whether the soil in a lake's drainage basin is rich or poor in nutrients. For this reason, thousands of lakes in a nutrient-poor belt stretching from northern Minnesota and Ontario to Maine and Quebec have changed very little since the last glacier retreated 10,000 years ago. Found in sandy or granitic terrain and unspoiled by man, these deep lakes have kept their gin-clear purity and their youth—they are oligotrophic. However, should we disturb their basins and fertilize their waters, they would immediately undergo cultural eutrophication.

Seepage from the septic tanks of just a few summer cabins and resorts, for example, is rich enough in nutrients to speed up the aging process. Cochran Lake, once a pure gem set in the northern Wisconsin wilds, has deteriorated so rapidly since the first of seven cottages was built on its shores ten years ago that it now looks like a 300-acre caldron of pea soup.

Lakes with large drainage basins in limestone terrain, which is usually high in phosphorus, are far more likely to show their age than lakes in granite or sandy basins. Because of their high nutrient content, they become shallow and die under encroaching cattails, reeds, and marsh grasses. But even though such lakes are already highly fertilized through natural means, whenever man is present his actions become the determining factor in the rate at which lakes fill in. Unless septic tank seepage is stanched, Cochran Lake—and thousands of other lakes in the resort areas of northern United States, Canada, Scandinavia, and the Alps—will be ruined.

While faulty septic tanks are a major source of nutrients in rural regions, their over-all contribution does not approach that of municipal sewage treatment plants. It has been estimated that 260 million pounds of phosphorus and 511 million pounds of nitrogen reach the nation's surface waters in the form of municipal sewage each year. Substantial amounts of these plant nutrients are discharged in effluent even after sewage is treated. As much as 75 per cent of the phosphorus in sewage comes from detergents. In addition to phosphorus and nitrogen, vitamins, amino acids, and growth hormones have been found in effluent—substances which contribute to the growth of algae and weeds. These growth stimulants are synthesized inadvertently in the biological processes of sewage treatment.

Since cities and villages across the country are rather impartial in dumping their sewage, rivers too are receiving their share of nutrients for eutrophication. This is most obvious wherever rivers have been dammed and currents slowed enough to give algae a chance to multiply. Because most of the nation's rivers have been systematically converted into series of impoundments since the 1930's, cultural eutrophication in these man-made lakes is now a common problem.

Algae blooms foul many of the man-made lakes in the TVA system, as well as the backwaters of hydroelectric dams on the Missouri and Snake rivers. The Mississippi—"father of waters"—carries partially treated sewage from Minneapolis and St. Paul forty miles downstream to Lake Pepin where algae thrive on sewage nutrients. As algae die and settle to the bottom with other sediments, the sluggish lake becomes shallower, increasing the possibility of spring flooding at Winona, Minn., and other downriver towns.

Formidable as it is, cultural eutrophication isn't inevitable. For example, five years ago officials from 70-plus local government bodies became alarmed about algal slime at sewage outlets on Lake Tahoe on the California-Nevada border, touted as one of the clearest lakes of North America. With the entire Tahoe basin's livelihood—tourism and gambling—at stake, they banded together and consulted sanitary engineers and limnologists. The prescribed cure: sewage treatment and diversion.

The South Tahoe Public Utility District, spurred by action committees of panicked businessmen, obtained a $10 million grant from Washington, raised another $9 million, and this year built a plant to treat sewage and a pipeline to carry effluent over a mountain pass to an irrigation reservoir. In diverting the effluent from the lake basin, the South Tahoe district provided California desert farms with nutrient-rich water. Meanwhile, the 42 voters of Round Hill, Nev., (most being gaming-house owners) floated a $5.8 million bond for their own sewage treatment-diversion project, fearing that smelly scum would hurt business at their lakeshore casinos and hotels. And if other Tahoe communities act right now, eutrophication should never smirch the lake's reputation for clarity—at least not for several thousands of years.

Diversion is currently the most effective measure we have to combat eutrophication, since it immediately deprives nuisance vegetation of nitrogen, phosphrous, and other nutrients. Moreover, until some technological breakthrough occurs in sewage treatment—now far from a completely effective process—diversion will continue to be the logical approach.

Sanitary engineers haven't yet found a process, chemical or biochemical, that offers an economical alternative to diverting effluent around lakes. Nevertheless, research is being pressed throughout the country. A pilot plant, for example, has been built to test a chemical process at Ely, Minn., the launching point for canoe trips into the Superior National Forest. Research teams have investigated the feasibility of using algae to remove nutrients from effluent. The plan is to grow algae in effluent-filled ponds until they have consumed all the nutrients. Then the algae would be harvested and the nutrient-poor effluent discharged directly into a lake, obviating the need for diversion. The researchers, in practice, have been stymied by several technical problems: for one, how can algae be put to use? Unfortunately, there is no demand for algae as a fertilizer or as an animal food.

While the logic of diverting effluent is obvious, particularly in light of the shortcomings of sewage treatment, it often hasn't been accepted. Opponents of diversion in Madison, Wis., for instance, argued that run-off from manured farmland—not municipal sewage—was the chief cause of eutrophication in Lake Monona, one of four lakes linked by the sluggish Yahara River. Nevertheless, soon after Madison began diverting all of its effluent around Lake Monona in 1936, the amount of copper sulfate needed to kill algae dropped from carloads to a few handfuls by 1956.

As Madison learned in the late 1930's, diversion requires a sense of responsibility and a regard for down-stream communities. That city, in sparing Lake Monona, poured its effluent into Lake Waubesa and indirectly into Lake Kegonsa. This half-measure promptly fertilized their waters. Residents, enraged by the proliferating algae that threatened to ruin their lakes, sued the city. In 1941 legislators passed an antipollution bill aimed at Madison. Gov. Julius Heil, however, vetoed it on the ground that there was no concensus on the main cause of eutrophication. Although the bill was later enacted, it was not until 1959 that Madison's effluent was diverted completely from Waubesa and Kegonsa.

Foes of diverting sewage plant effluent around Seattle's Lake Washington to Puget Sound claimed that farm runoff was responsible for cultural eutrophication—just as had been claimed in Madison—even though the 24-mile long lake was absorbing the effluent of ten treatment plants by 1950. In this case, it took the lavender bloom of

Oscillatoria rubescens, the same alga that discolored Lake Zürich, to awaken the public. Mass apathy dissolved, but a welter of conflicting opinions took its place and for a time threatened to defeat the attempts to divert sewage around Lake Washington. Eventually, 5,000 women, backed by the Municipal League and the League of Women Voters, went from door to door, mustering support for a bond issue to be used for diversion. Recently, voters reversed the stand they took in 1956 and approved the bond issue and a charter for a new metropolitan governmental body that would cope with the sewage problem. Although the diversion project hasn't been completed, the picturesque lake is nearly as clear as it was eighteen years ago. Its clarity should continue to improve as its effluent is diverted to the tide-washed sound.

Diversion projects are sometimes condemned as costly boondoggles, since rainfall inevitably contributes so much nitrogen to lakes, whether or not diversion is practiced. Critics, however, fail to consider, or choose to ignore, the fact that municipal sewage—not rainfall—is the major source of phosphorous, the other stimulant of cultural eutrophication.

There is no standard remedy for ailing lakes. Diversion is no cure-all. It has disadvantages in addition to its expense and capacity for creating or aggravating water problems elsewhere. Diversion can negate itself, particularly in water-scarce regions where sources of clean water that might have flushed lakes clean have been channeled instead into diversion systems to carry away sewage effluent. Thus, the advantage of diverting effluent is sometimes offset by the disadvantage of diverting potentially cleansing inflows. Engineers, aware of this drawback, are exploring the possibility of using "infertile" water from the Columbia River to rinse Moses Lake in central Washington.

Harvesting weeds and fish to remove nutrients from fertilized lakes promises to supplement diversion projects very effectively. Seining crews, for example, hauled 40 pounds of carp per acre from Madison's Lake Mendota. Since the nitrogen and phosphorus content of fish flesh is 2.5 and 0.2 per cent, respectively, 500 pounds of carp harvested removed 12.5 pounds of nitrogen and a pound of phosphorus from the lakes. The 1966 carp catch yielded 16,000 pounds of nitrogen and 800 pounds of phosphorus. The nutrient "yield" of this method could be easily expanded in most eutrophic lakes with more intensive harvesting of carp and other rough fish.

A program to perfect weed-harvesting techniques and to evaluate their effectiveness was begun last summer on Lake Sallie, a densely populated lake in the vacation area of Detroit Lakes, Minn. By "cropping" one-third of the lake three times to a depth of five feet below the surface, the researchers expect to remove substantially more nutrients in one summer than enter the lake in a year. The outlook for harvesting algae on a large scale, however, isn't nearly as bright. No economical equipment has been developed to skim algae off lakes and impoundments.

While inventors are working on a variety of devices that "suck up algae like vacuum cleaners" or "strain algal scum from lakes," members of the National Eutrophication Research Program are hunting for a herbivore, perhaps some water flea or fish, with a huge appetite for algae. The search may be extended abroad if no algae-craving counterparts can be found to the weed-eating snails used in the southeastern United States. Precautions, however, must be taken to screen out "exotic" species that could become nuisances. Also, of two viruses known to cause diseases among blue-green algae, one is being tested, and more virulent species are being sought.

The best that can be said for spraying chemical poisons on lakes in the grip of algae and weeds is that it is usually a futile undertaking. Treating a lake with copper sulfate or other toxic chemicals is no more effective than taking aspirin for a brain tumor. It offers only temporary relief, masking the symptoms of cultural eutrophication. In the long run it makes a lake sicker. Poisoning algae and weeds simply accelerates the natural process of growth, death, and decay, thereby freeing nutrients for another cycle of plant production. This has been borne out by studies of blue-green algae in Oregon's Upper Klamath Lake, which first showed signs of eutrophication eighty years ago. In the peak growth month of August, Klamath algae contain within their cells three times as much nitrogen as is dissolved in the lake. Killing an alga therefore would release this nitrogen for further growth.

Chemical poisons should be used only as a last resort, for once dumped into a body of water, they cannot be confined to one locality. They dissolve and spread far beyond the area treated. Dispersed eventually by wave action throughout a lake, they adversely alter the fragile fabric of aquatic communities of many species. Too little is known about the sublethal effects of such poisons to risk their use.

We can take several steps—all more effective than chemical poisoning—to bolster the two-pronged attack of diverting effluent and removing nutrients. By separating sewer systems for storm runoff and sewage, communities can forestall the lake fertilization that occurs when nutrient-rich combined sewers overflow after heavy rains. Damaging septic tank seepage can be cured by passing legislation that would require that tanks be installed at a "safe" distance from bodies of water. Wisconsin legislators, for instance, recently decided that lakes and rivers would be safe from fertilization only if septic tanks are set back as far as

1,000 feet and 300 feet, respectively. They also set more stringent specifications for septic tanks themselves. The specifications vary with the porosity of the soil.

Livestock growers could also fight cultural eutrophication by adopting the European practice of fluidizing and storing manure in vats, from winter freeze-up until the spring thaw, and then spraying it on their fields when the soil can better absorb it. The bulk of the manure now spread on frozen cropland runs off with spring meltwater to fertilize lakes and streams. This is significant considering the fact that the amount of manure, produced each year in the Midwest alone, is equivalent to human sewage of a population of 350 million. The chief drawback of the European method is its expense. Federal funds for water and soil conservation, however, could possibly be used to help farmers pay for costly storage tanks and spraying equipment.

We still have much to learn about the complexities of the eutrophication process. There is an urgent need for more research on such factors as a lake's depth, size, and configuration, its drainage basin and sources of nutrients, and its aquatic communities. But such research is complicated because biological and ecological factors vary with every lake, pond, and reservoir. Systems analysis with computers promises to be the research tool that will enable us to supply precise data on the potentialities of eutrophication.

Improved research, however, is only part of the solution to this dilemma. Unfortunately there isn't time to raise an enlightened generation to cope with eutrophication. Although we don't have all the answers, we must heed the warnings given by the waterways around us that have fallen into an advanced state of decay. If we delay, the price may be too dear to pay.

12

"Arctic Walk"

John P. Milton

"A bit of a dream, a portion of persistence, and an unknown quantity of luck" began John P. Milton's expedition through the Arctic Wildlife Range in Alaska. His involvement with the natural environment has taken Mr. Milton through most of the American continent, parts of Africa, and Europe. The author of nmerous books relating to natural resources, he is currently developing a book series celebrating the earth's last "wild places" and the measures needed to preserve them. Mr. Milton is Deputy Director of the International Programs Divisions of The Conservation Foundation.

Between July 22 and August 29, 1967, Kenneth Brower, Steven Pearson, and I walked from the southern foothills of Alaska's Brooks Range, across the high mountains of the Arctic Divide, then down across the open tundra to the Barter Island DEW-line station, on the Arctic Ocean—a transect, from south to north, across the Arctic Wildlife Range. Much of our route had no record of earlier exploration. Excerpts from my journal of this trip are contained in the following pages.

On this 300-mile walk, my own purpose was to blend into the soul of these Arctic mountains and plains, to sense the moods and manifestations of the land. Our journey was a lesson in solitude, a vanishing experience in an age when most of mankind is crowded, yet lonely.

In July, 1968, a year after our walk, the Atlantic Richfield Company and Humble Oil and Refining Company announced the discovery of "one of the largest petroleum accumulations known to the world today" along the Arctic slope of the Brooks Range. This field may be equal to from one-tenth to one-fifth of the oil reserves of the entire North American continent. Don Randall, an oil expert and geologist, commented shortly after the announcement: "This discovery . . . will open the door to a vast and fabulous play stretching from Alaska across the Canadian Arctic."

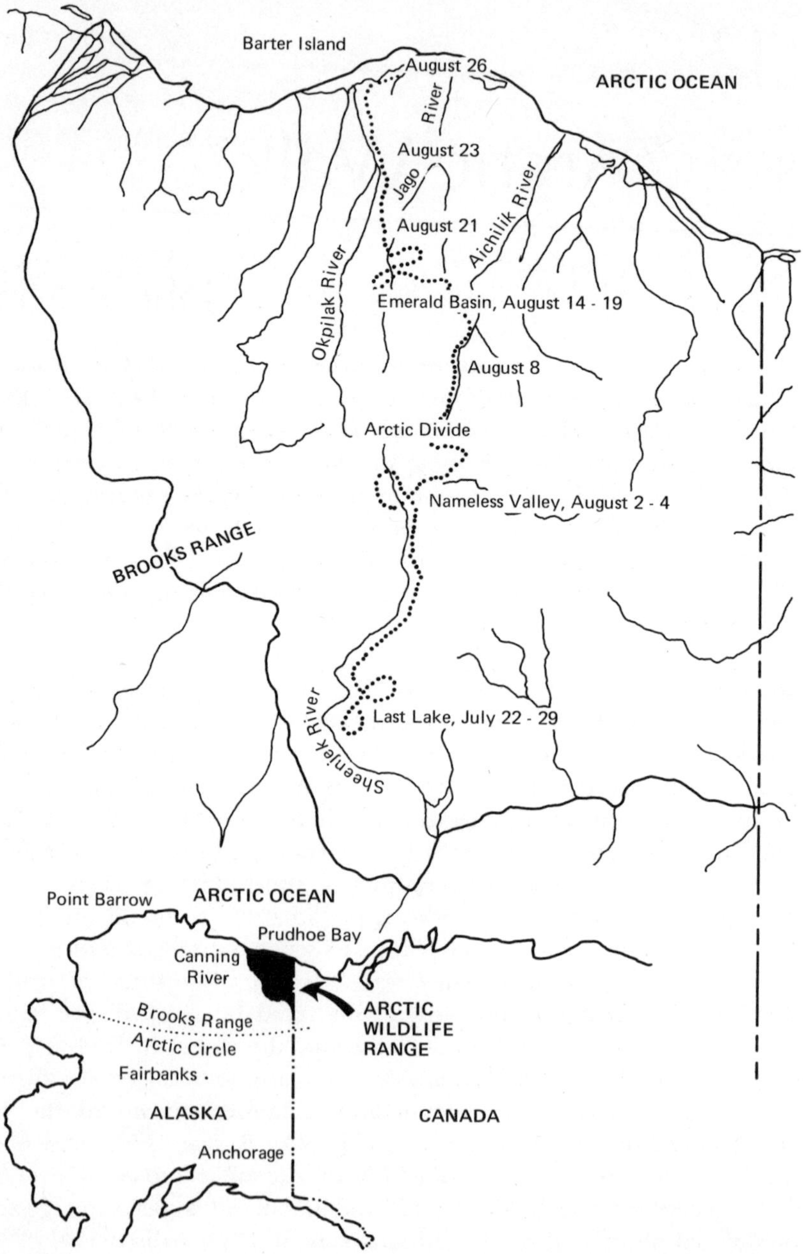
Barter Island
August 26
ARCTIC OCEAN
River
August 23
Jago
Aichilik River
August 21
Okpilak River
Emerald Basin, August 14 - 19
August 8
Arctic Divide
Nameless Valley, August 2 - 4
BROOKS RANGE
Sheenjiek River
Last Lake, July 22 - 29
Point Barrow
ARCTIC OCEAN
Prudhoe Bay
Canning River
ARCTIC WILDLIFE RANGE
Brooks Range
Arctic Circle
Fairbanks
ALASKA
CANADA
Anchorage

FIGURE 12-1

Over the months since the discovery, a fever of activity has spread throughout the North American Arctic, particularly concentrated at Prudhoe Bay—located on the Arctic coast about 300 miles north of Fairbanks and 150 miles southeast of Point Barrow. It is here that the major strike was made and where most of the drilling is being done at present; over much of the rest of the Arctic, plans for new exploration are rapidly forming. On September 23, 1968, *The New York Times* reported that recent seismic surveys have identified several good prospects for drilling in Canada's high Arctic. They also stated: "In a year or two, American and Canadian oil companies may pour survey and drilling crews into the northwestern corner of Canada." The play has begun. The latest rush for black gold is on.

Climatic and transportation problems, however, are already posing some difficulties. In Alaska there are no roads or major ports north of the 600-mile Brooks Range, which runs across the northern half of the state. A decision has already been made to run one pipeline over these mountains to the southern coast of Alaska. There also have been proposals for a second pipeline to cross from Prudhoe Bay into Canada's Mackenzie Delta region, where oil could then be shipped up the Mackenzie River or through the Northwest Passage to the eastern United States. This would allow oil companies to avoid the restrictions of the Jones Act, which requires that all interstate shipping must be on U.S. carriers. A port in Canada would allow use of less expensive foreign shipping.

In 1960, Secretary of the Interior Seaton established the 9-million-acre Arctic Wildlife Range in the rugged eastern Brooks Range. In size, the protected area is slightly larger than the state of Vermont. The reserve extends approximately 140 miles from the Arctic coast in the north to the foothills of the Brooks Range in the south. From the Canadian border forming its eastern boundary, the reserve runs 120 miles west to Alaska's Canning River. Only fifty miles west of the Canning River lies Prudhoe Bay and the major locus of the oil discovery.

The Arctic Wildlife Range is the largest single federal biological reserve in the country. It is the only protected wilderness area in the mountains and tundra of Arctic Alaska; the home of moose, lynx, hares, caribou, polar and grizzly bears, Dall sheep, wolves, wolverines, Arctic foxes, and many other animals. The necessarily large size of the reserve has helped insure that this virtually undisturbed portion of Alaska has a high degree of biological self-sufficiency. All the major habitats typical of northern Alaska are found within its boundaries: taiga; Arctic mountains, rivers and glaciers; open tundra; and the ecotone between tundra and taiga.

Wilderness protection in the Arctic Wildlife Range may be endangered by the new oil discoveries at Prudhoe Bay. Should the Secretary of the Interior so desire, he could open the reserve to exploration and drilling. This would cause severe environmental damage, since, due to the Arctic's climate and abbreviated growing season, surface scars on the tundra, from only the occasional passing of vehicles, remain for dozens of years and often cause a cycle of accelerated permafrost melting and erosion. If the proposed pipeline from Prudhoe Bay into Arctic Canada is built, it would cut across the open tundra of the Arctic Wildlife Range to compound the damage done by exploration and drilling.

Dumping of oil drums by oil exploration parties has already marred the beauty of many parts of Arctic Alaska. Similar activity in the wildlife range would quickly shatter its wilderness mood, as would roads, camps, pipelines, oil rigs, buildings, and other repulsive accouterments of oil technology. On the open plains of the Brooks Range's north slope, even minor irregularities in the landscape are visible for greater distances. The disruptive influence of a single tower, for example, would extend over an area of many hundred square miles.

A fire hazard could also arise, particularly along the range's south slope. A fire that seriously damaged the region's lichens would prove particularly devastating to caribou. These plants are the basic food for caribou and often require up to 120 years to regenerate after burning.

There is still another danger to both terrestrial and marine ecosystems posed by problems of oil pollution. Inevitably there will be problems of liquid waste disposal into rivers and marine waters, particularly if any refining is done along the Arctic coast. The greater pollution threat, however, comes from the potential danger of oil spills from tankers or leakage from any offshore wells that may be drilled. A disaster like the wreck of the *Torrey Canyon* or the recent Santa Barbara oil leak would have severe effects along the Arctic shores of Alaska and Canada. During the few months when these coasts are relatively free of ice, the main ice packs are never far from shore, leaving only narrow strips of open water. An Arctic oil slick would be concentrated along the shoreline by the ice pack, allowing little chance for dispersal, and the cold water might further inhibit the activity of microorganisms important in breaking down the oil. The impact on the Arctic's rich marine plankton, fish, and mammals, and upon the extensive summer habitat of many migratory waterfowl, would be tremendously detrimental. Also, what would be the impact on the Eskimo's way of life if the whales, seals, fish, and birds upon which they depend are depleted?

In the current enthusiasm over oil in the Arctic, little, if any, attention is being given to these major environmental problems.

JOHN P. MILTON
March, 1969

Friday, July 21, 1967

After three days of rain and trip preparation in Fairbanks, we are about to leave for Last Lake near the upper Sheenjek River.

Our 250 pounds of gear are now loaded in the little plane that will carry us to our remote first campsite—Last Lake. For the next month or so—hopefully—we will see no other men. We will be alone, on our 300-mile hike, to take in the wilderness of the Arctic Wildlife Range. To our knowledge, no other men have made this south-north journey through the high mountain passes of the Brooks Range. From now on, our life will center on the immediate horizon. No media, no middlemen.

Saturday, July 22

We are here. My sense of space and distance are greatly expanded by this country. It is incredibly beautiful. The broad valley of the Sheenjek curves through a tumbled series of steep side valleys and slate-grey mountains practically barren of visible vegetation. The valley floor is green and lush, with numerous patches of open forest breaking up the grassland and wet muskeg. To the north, at the head of the Sheenjek, a band of snowy mountains reaches skyward between 6,000 and 7,000 feet. (At Last Lake the valley floor has an elevation of 2,500 feet.) Over all this, a wonderfully clear atmosphere lets the eye wander on and on, drawing it deeper into the hidden mystery of far valleys and mountain passes. The sun, revolving slowly in a great circle, casts a deep, warm glow over the wilderness. For only a few hours it disappears behind the northern mountains, casting a fiery sunset all "night" long. I cannot walk far in any direction without having to stop and look about me. We will camp here for several days.

Saturday, July 29

It's nearly midnight and I can still write by the evening light. We find abundant signs of wolf, lynx, Dall sheep, moose, grizzly, caribou, and smaller animals here. Ken reported seeing a yellow-haired porcupine that looked "just like a small bear. When I talked to him he faced me and wasn't afraid. When I came back for a second look, he grew suspicious and waved his spiny tail at me."

Ken and I sat until late in the evening watching and photographing the present two rulers of a little kingdom of beavers upstream. Probably the northernmost colony of beavers in this hemisphere. They splashed and whacked their tails noisily when we first arrived, but after that they seemed to quiet down. We sat still. The two beavers sliced wakes of liquid silver as they churned to and fro in front of us. Just once, the larger one—the male—overcame his fear enough to come almost to our feet. He lay still, flat tail slightly waving, peering up at us out of suspicious old-maid's eyes. Then, *whack,* and he was gone.

Yesterday, I walked south a mile to frozen "Ice Lake." The lake is cut across by a flow of clear water, which slices a serpentine course through bluish-white ice walls from eight to ten feet high. A small ice cave sheltered me for a few minutes when a storm passed over. The ice banks reached over me—not quite meeting. The inner walls, drizzling in their own steadier rhythm than the storm above, were neatly scalloped—as though by some fine craftsman's chisel. After the storm was over I left and plodded back across the brownish moorland to camp.

Friday, August 4

Today, exploring side valleys, we came to a very high, impassable cliff and talus slope that dropped steeply into a tributary of the Sheenjek. We had to climb 800 to 1,000 feet over it to get by. But the view from the top was superb. The valley, lined by impressive ridges and peaks, curved slowly to the west. A river wound down the center, lined with gravel bars and willows gleaming from a patch of clear sunlight here and there. Up from the stream, the green valley bottom sloped gently to the brown grasslands and gray rock of the mountainsides. The whole valley seemed very special and secret. The knowledge that we were probably the first white men to set foot in it only added to the impression of a little world set apart.

These mountain valleys seem to have a strong personality, an individuality that sticks in the mind. And so it is here, in Nameless Valley. Arctic mountains are distinguished by an ever-present mood of darkness, cold, and snow. Winter is a constant, though usually hidden, companion during the short Arctic summer; the quickness of the summer only intensifies an awareness that here warmth is only a brief respite from cold, that light will soon be followed by a deep and much longer-lasting darkness. This mood dominates the land and every living thing in it. The mountains also know the cold of glaciers; the ice wedges carve their slopes and chip away the cliffs, leaving great cones of talus to pile up below. Summer is but a short, quiet interruption—a convenient punctuation to one more dark and wintry cycle.

Walking back to camp along a high ridge of rock in the valley center, I was struck by the quality of this wilderness: Tiny details like the mats of red moss mixed with green carpeting: little mountain rivulets and the larger scale of grizzled mountain slopes; the vast expanses of wide valley; far, snow-flecked peaks, clouds, and sky—all are contributing elements. This is wilderness on a scale that the mountain men knew in our far west during earlier days. To have the mood requires hundreds of miles of empty land and large expanses of unexplored territory. This Brooks Range wilderness still has these elements in abundance, and imposes upon you a need for self-reliance.

Here is an atmosphere of nature at its untamed, uncivilized best. The wilderness stands on its own: free, not propped by access roads, park rangers, interpretive centers, and regulations on use, as in the quasi-wilderness of our national parks. Here there is no prostitution of the freedom so essential to wilderness—and the quality of the experience reflects this.

Tuesday, August 8

The hour is 10:00 P.M. and it is snowing steadily. We are camped at the junction of the two main headwater streams that combine to form the upper Aichilik River.

By five today we were all chilled, with freezing cold feet and hands. We set up camp in record time and stripping off our wet pants, followed by a change into warm socks, we plunged into the double cocoon of tent and sleeping bag. Dinner—freeze-dried pork chops and mixed sliced potatoes and vegetables—was cooked over the little portable stove in Steve's tent. That little stove has been doing very well by us ever since we crossed over the Arctic Divide; recently we've had no opportunity to build a wood fire (either no fuel or too wet) although today river willows reappeared, and had the weather been different, we might have. The stove has been burning for well over an hour on one filling (instead of the prescribed 50 minutes). After so many "4-man meal" food packages that would barely serve two, it's good to have some surplus at least with our cooking fuel.

My injured ankle was a little better today, although for the last two hours I was in almost constant pain while walking. I've noticed I prefer to walk either last or first and to have our group well spaced out. When in a tight group, I always tend to feel distracted by the sounds and movements of the others. Much better to wander as individuals, collecting our own feelings and experiences to be shared in the evening or at rest stops.

Saturday, August 12

In spite of Steve's and my ankle problems, we managed to cover another eight miles today between 2:00 and 7:00 P.M. The northern border of the Brooks Range is now quite near—and beyond that, open tundra.

Somehow, the tundra now seems to represent the major unknown for us. We worry about crossing the Jago River, but much more about traversing the tundra. Will blizzards hit? High winds? Is there enough fuel? Can we be sure of finding the right stream to follow in all that flatness? How will we know when to cut cross-country to Barter Island—lacking landmarks? Will the walking be all muskeg, tussock grass, and bog—or gravel bar and level grassland? Will our feet, legs, and general health hold out long enough to get us there? These are some of the questions we've asked oursleves while pondering the tundra. And so it must always be with any unknown challenge. Such questions are the springs of adventure.

An hour ago Ken strode into camp after some early morning hiking. From an overlook he spotted a medium-sized grizzly that came within 70 yards of camp, rose on his hind legs, then apparently caught our scent. He bolted up into a side valley, running continuously until out of sight. What is it that makes such a powerful beast run? Fear of the unknown? We are surely the first men he has ever smelled; but if he has no other animal to fear in his domain, why run from a new smell? No other animal can conquer the grizzly in his Arctic home, and it should be outside his experience to have to take flight. But he ran.

It is now nearly midnight, and I write by candlelight; the horizon to the north outside my tent is afire with a deep, reddish glow. The thermometer stands at 23° F. Inside the tent it is a warm 28 degrees.

Monday, August 14

Today has been magnificent. We've walked about eight miles, over two passes, and camped atop a wide, two- to three-thousand-foot-high plateau (which we call Emerald Basin) at the northern edge of the Brooks Range, near the junction of the mountains and the barren tundra stretching north to the Arctic Ocean.

On a mud bar in Green Willow Creek, I passed a line of wolf tracks, unusually large in size. Once they met and mingled with caribou tracks of about the same age. An encounter? What a free-ranging life for wolf and caribou up here; they are animals needing broad stretches of wilderness to survive—as did once the bison and plains wolf of our long-lost western frontier. In his inability to compromise with civilization, the wolf is almost entirely gone from the contiguous U.S. But here is the old wolf range on a vast scale, as before.

Earlier in the evening, a red sun settled down above Lone Mountain, and fog sifted in thickly from the tundra through a pass and sent out numerous fingers of white across the valley floor. The sun turned the top of the fog to gold and bathed the green valley in amber light. Sheer, snowy mountains looming over the scene turned rosy colored from flanks to summits, while ice formed in thin sheets of transparent crystal on the small pools between tussock grass. All the agony of injured foot and blizzard were worth these few hours of evening.

This land seems to be forever unfolding new surprises, both in weather and terrain. These two elements combine and recombine in an infinite variety of form and light. Who can say they have seen these Arctic mountains as they really are?—they are too changing and too varied. Each of us who has come here has lived but a part of the truth; tomorrow will not be the same.

Tuesday, August 15

About an hour after rising in the morning, which dawned clear and bright, Steve yelled "caribou!" I grabbed my binoculars, camera, the two telephotos, and was out of the tent. About a half mile away, two bulls lay contentedly in the snow. While Ken watched, Steve and I began to stalk them, got to within one-quarter mile of them, then sat down to watch and wait.

Slowly, the bulls got up from their resting place and grazed for a short while between snow patches. Then, as though curious to see what we were, they began to trot directly toward us. Our cameras clicked madly. They passed within 30 yards, heads lifted proudly and their antlers still in velvet. After passing us they nearly went through camp—20 feet away from Ken then on out into Emerald Basin.

Friday, August 18

I now feel little desire to change my environment from Arctic wilderness to civilization. All of us have now been here long enough to settle completely into our new pattern of life. Although difficult, it has its compensatory comforts and the sharp joys of wilderness experience. At each of our three main base camps, Last Lake, Nameless Valley, and now here at the mouth of the Jago's mountain valley near the beginning of the tundra, this sense of belonging, of a place defining our normal life, has been particularly strong. But, during those times when we've been on the move day after day, then that nomadism seems to become the normal pattern of existence. How quickly man adapts.

And yet, over periods of time, deficiencies in life appear—in either the wholly civilized life or a life constantly in the wilderness. It is a life spent contrasting and living alternately in both worlds that seems best to me. Indeed, the very words themselves, "civilization" and "wilderness," require each others' existence to have real meaning.

When we go into wilderness we are consciously reversing a historical process whereby civilization has gradually acquired more and more power to manipulate and change the earth to serve man's own ends. But, as in the case of the "ring of power" in Tolkien's classic trilogy, power over nature carries with it a new curse; the newly found power more and more begins to control us and limit our choices. Despite the many new freedoms that advancing technology has brought us, application of each new advance involves us in an ever deepening web of causes and consequences that require considerable new limitations on older freedoms if we are to sustain the new ones.

Saturday, August 19

Today we climbed up to a long, narrow lake just west of the Jago River. An oil exploration crew had been there before us. The beauty of this lake is now diminished by the presence of over 20 scattered oil drums; the tattered ruins of a large, domed shelter; and litter of all kinds strewn about. Man's insensitivity to a wilderness landscape can be appalling. These scars will remain for a generation or more—yet the wreckage was littered about with no more thought for those who might follow here, than a group of monkeys let loose in a fine French restaurant would feel for the next day's patrons.

Fortune has provided for us in one way, however, for amid the litter we found tins of cooking oil and sugar and a pancake turner to allow us to turn out whole cakes. We also found a new frying pan handle to replace the one lost and a new teapot top to fill in for our badly battered one. We'd lost two forks and a cup—those also we discovered and took. Finally, I added a fourth plate as a luxury item to place our little stove on or to cut food in. All things considered, we had good scavenging. But I would rather forgo the extras and find a clean tundra lake.

Monday, August 21

After about two hours of walking this morning, the fog gradually and steadily lifted. Abruptly the tundra was transformed as the sun broke through to turn the earth to gold. It is almost impossible to describe the vastness and special beauty of this open tundra. I was struck by the absolute flatness of the northern horizon—we were able to see over fifty miles to where the coastal clouds began. On either side the warm, rich tone of the tundra rolled out in long, gradual waves of gentle slopes followed by broad basins. Behind us, to the south, the brown plains sweep up to the very edge of the granite peaks of the Brooks Range, decked with long white glaciers. The subtle shadings and gentle contours of the prairielike tundra seem strangely accentuated by the sharp, savage mountain topography and color. And tonight, another mood comes over the tundra as the sun sets and the land colors a deep burgundy.

This trip has helped sort out my existence. When I'm caught up in the day-by-day turmoil of our complex society, it becomes difficult to separate the meaningful from the meaningless. On this expedition many of my problems—or what I saw as problems—began to take on new form and perspective. The relative simplicity of wilderness life cleanses you of numerous petty concerns that are so typical of our waking time in the civilized world. Out here goals are basic and obvious: good health, warm shelter, food, a fire, where to camp, where to walk, the enjoyment of your surroundings.

Wednesday, August 23

It is midmorning as I write in camp, and the sun is attempting to burn through the clouds. We will continue along the Okpilak River toward the Arctic Ocean. How simply and easily we have moved through this wilderness—and what a contrast to (I like to think) the cumbersome, ugly movements of the oilmen who tried to bring all of civilized life here with them, and succeeded most in scarring the land with oil drums: the symbol of their particular rape of the land. In the wilderness, man should move like a nomad, with the aim of leaving the least possible traces of his passing behind him. Here is the opposite ethic to Ozymandias: man is measured more by what does *not* remain, than by what does.

We have done well in abiding by this ethic, I feel; bits of paper are burned, our few small cans flattened and buried, our campfire stones scattered. All that remains is a patch of slightly flattened moss and grass, no more than a small herd of caribou or family of wolves would leave behind following a long rest.

Saturday, August 26

Morning came with a cold, damp fog still hanging over our camp. We ate a full breakfast of eggs mixed with what was left of last night's vegetables, and a meat and bacon bar. By the time we were through, the fog had lifted and the sun was shining. At 1:00 P.M. we broke camp and headed north. About two miles north of camp, we topped a rise—the Arctic Ocean lay before us!

Sandwiched between bright blue sky and sea, the brown tundra ended suddenly in a series of low cliffs running down the coast as far as we could see. The creek widened into a small marsh and beach. We followed it. A series of barrier beaches and islands, the largest of which is Barter (site of a U.S. DEW-line station), lay from a few hundred yards to several miles off the coast.

Far off to the north, the brilliant white of several widely spaced icebergs shimmered on the horizon. Around us, the Arctic tundra glowed golden in the sunlight of late afternoon. At the edge of several small marsh ponds, fresh grizzly tracks showed in the mud. And dead ahead, the unreal twin radar towers of DEW-line drew our sight and steps toward them. Fifty miles behind us, the Brooks Range front threw a solid wall of snow-covered summits into the southern skies. Had we really walked over them and across the intervening distance? The view dazzled us, and a floating sensation began to take possession of my powers of awareness—a feeling that grew, as Barter's incredible pile of space age and war technology closed around us, as if to insulate us not only against some obscure threat from Siberia but also from the more real threat to civilization implicit in the vastness of this surrounding wilderness.

After crossing the channel to Barter Island with the aid of an Eskimo boatman, we finally found the military commander and a civilian aide. "Walked over the Brooks Range? . . . over the high peak region? . . . for God's sake—why?"

13

"The Florence Floods"

by Richard M. Klein

In "The Florence Floods" Richard M. Klein gives a plant biologist's analysis of the 1966 Italian disaster.

As it did to other presumably educated Americans, the announcement of the Florentine flood of November 4, 1966, aroused in me the immediate reaction of horror felt by most people when they hear of a natural catastrophe. Fragmentary stories on the wire services were followed by detailed descriptions of specific paintings damaged or destroyed, books soaked with water and oil, and accounts of death, hardship, and personal heroism. I shuddered at the evoked word picture of the odors of sewage, rotting animal and human flesh, mold and mud and oil. In the succeeding days, as is usual in our overcommunicated world, the stories receded into the back pages of the newspapers and fresh stories of other tragedies took over my attention.

Yet, there remained a nagging feeling that this flood should never have happened at the level of intensity and with such severe loss of the treasures of Renaissance Italy—indeed of Western civilization. As I read more about the history of the Florentine region, I began to realize that the 1966 flood was an almost classic example of how to ensure that floods will recur.

The area known as Tuscany is a well-marked and fairly compact geological unit, which rises gradually from the western coast of Italy to the heights of the Apennine Mountains. It is a harsh land of hills and valleys, of winter rains and summer droughts. The valley of the Arno lies on the northern margin of the African trade winds and on the southern margin of the prevailing westerlies that bring rain from the Atlantic Ocean. In winter, the warm air above the sea forms a low-pressure cap that draws winds over the water. As these warm, moisture-laden winds flow up the hills to the Casentino mountain range, they cool, and the rain may fall furiously for several days in succession. Of the twenty-five to forty inches of rainfall that Tuscany receives annually, over 80 percent falls between October and January—much of this between late October through the middle of December. In the summer, there may be no rain for several months as the hot, rapidly moving winds from Algeria simply do not pick up enough moisture from the sea to supply rain.

Today, the hills surrounding Florence can scarcely support olives and grapes. There are occasional patches of scrub, including lavender, myrtle, rosemary, and thyme: aromatic, romantic, lovely to look at in flower, and completely useless as ground cover. The soil is leached-out clay with pockets of stone and sand. The sun beats down on impoverished and badly worn soil, changing it into a cementlike hardpan. In spite of, or perhaps because of, its aridity, the land above the city of Florence has a very real beauty. From the surrounding hills, one can take in the peaks of the Casentino mountain range, bared and gleaming in the summer sun. First in abrupt descents and then more gently, the highlands fall away to the foothills. Sere, brown, and rocky, the hillsides catch the early morning or late afternoon sun, leaving the hollows in deep shadow. From a distance, it seems a landscape devoid of plants.

The present city of Florence stands on land that has been a part of the civilized world for more than 3,000 years. Excavations have unearthed artifacts dating back to the Bronze Age, and there is some evidence that neolithic cultures flourished there even earlier. And well they might have, for the Tuscan valleys and hills were fertile and wooded, the climate then was generally mild, with just enough seasonal variation to stimulate, but not debilitate, the people, and there was reasonable protection from marauders from the north. The major river, the Arno, was a freshwater stream, apparently abounding with fish, that flowed throughout the year and was broad enough to permit travel, and hence trade, with the coastal peoples through the delta that is now the city of Pisa.

The earliest complex civilization about which we have reasonable information is that of the Etruscans. Apparently they settled in what is now Tuscany about 800-600 B.C., coming from northwest Asia Minor via the Caucasus. In addition to being farming peoples, they were experienced merchants, and much of our knowledge of them is derived from their commercial activities. The Etruscans found, settled, and exploited a land of some natural beauty. The Arno and its tributaries were clean, clear, fishable, and navigable. The hills were completely forested. In the valleys were found the Aleppo pine and two evergreen oaks, the Holm oak and the Valonia oak. These formed fairly open forests with understories of various small trees and shrubs. The midmountains contained excellent stands of beech, Spanish chestnut, and deciduous oaks, and the upper mountains were closed with black pine and several species of fir. There was an abundance of animal life, for figures on Etruscan pottery depict deer and bear hunts. In the fourth century B.C., Theophrastus mentioned the beech groves of Tuscany as being of such excellent quality that the wood was bought from the Etruscans for the construction of ships' keels. Some records suggest that they also sold fir lumber to Carthage. Even by 300 B.C. the forests in the hills were so dense that the Roman legions hesitated to invade. The plant cover was still notable several hundred years later when Strabo reported that the central Apennines and the hills of Etruria provided wood for the construction of Rome. The Sila hill range yielded "fine trees" and pitch for shipbuilding activities centered near Pisa at the mouth of the Arno.

The Etruscans were a vigorous people who attempted to use their land for many activities related to commerce. They planted olive trees and grapevines on the hillsides and farmed the valleys for wheat and barley. They mined some copper, zinc, and tin to make bronze for tools and weapons, and they probably also smelted mercury and some iron. Sheep were grazed on the lower hills to supply the white, black, and brown wools that were much prized for their softness, fineness, and natural colors. Thus, some deforestation and grazing injured the land, but the populations were too small to severely damage the fragile hills. By the end of the Etruscan hegemony, the Romans were lavish in their praise of this still-wooded, fertile, and well-watered land.

The first adequate records of floods date from the twelfth century, actually 1117, when the Ponte Vecchio, then the only bridge over the Arno, was swept away. These records show clearly that the Tuscan hills were denuded, the Arno was a seasonal stream, and the land was plagued by drought. If we date the demise of Roman rule at about A.D. 300, and we know that the land was ruined by 1117, we are left with approximately 800 years to account for in terms of physical geography. What happened to the forests, the clear streams, the good soil, and the fine climate? Can we use the meager records to fill in the details of man's rape of this land?

The primary city of the Etruscans was not the area of present-day Florence, but was the now small hill town of Fiesole. Present-day Florence was a lower suburb about 15 miles downstream on the Arno. The plain of the Arno is broken into a succession of old lake beds, which filled with debris and in which beds of travertine were deposited. Florence itself was initially situated where the Arno issued from the mountains at about 180 feet above sea level, a few miles upstream from its present location. The upriver location was determined by a simple fact—the site of the present city was a marsh. Indeed, when the winter snows melted in February, the river was not easily forded. Livy noted that in the late winter of 217 B.C., Hannibal and his armies floundered for four days in the muck of the valley floor. Yet, the river bottomland was very fertile, producing an excellent crop of hard wheat, some millet, and good barley. The snow melt served to supply nutrients, and the water moved out to sea by March, leaving enough time to plant. There was a second reason for the lowly political position of Florence, one related to the defense of the area. The Etruscan military wanted its citadel on the heights, not on the nearly indefensible valley floor.

Fiesole, as the political and defense center, tended to gather the greater share of the population, but it soon became apparent that this small, hilly area could not provide an adequate living for the populace. The hills about Fiesole had to be intensively cultivated and the only practical way to do this was by terracing. The technique was certainly not invented by the Etruscans; the Greeks and other peoples of Asia Minor had been cutting horizontal ridges into their hills for eons. In the Tuscan context, terracing was necessary more as a means of fully utilizing all land within the protected reach of the fort, than as a conservation measure (although it certainly served well in this respect). Cereal crops were not planted on the terraces, but were grown on the Arno Valley floor, and the crop was stored in the granaries of Fiesole. Vines were planted on the terraces, particularly on the drier, stony, south-facing slopes of the hills. Typically, the vines were festooned from tree to tree. Initially, the trees were native species, but these were replaced by the olive prior to the advent of the Roman invasions from the south. Thus, even before the rule of Rome, some of the plant cover on the hills had been removed. Since the Romans left no records of flooding, we can assume that the major portion of the Tuscan hills was still covered with plants that sopped up the water and prevented the destructive run-off that resulted in floods.

Tuscany, being on the route from the lands of the Franks and the Goths to that of the Romans, served as a staging ground for many armies. The mountains were no longer an effective barrier from the north as logistics and disciplined armies replaced wandering bands of military plunderers. Records are hazy and contradictory, but it seems that the

Romans defeated the Etruscans several times between 300 and 200 B.C. and that about 200 or 150 B.C. Roman rule was firmly established. The Roman road system was extended through Tuscany, and during its construction the engineers tried to stay in the piedmont except where there was a necessary valley river crossing. The Via Julia Augusta crossed the Arno at the site of the present Ponte Vecchio to connect with the Via Cassia going to the north. Even then, the Roman legions knew it as a bad crossing because it was muddy and tended to flood in the spring. There are no records of massive winter floods, however.

Under the protection of the legions and the Roman governor, Sulla, the population of Fiesole began to move down the Arno to where life was easier. When Sulla was defeated by invading Franks, Florence was leveled and burned, and again the people retreated to the hills. By 59 B.C., Florence was restored under the auspices of Julius Caesar to its present location and given the name Julia Augusta Florentina, the latter word apparently being a good omen rather than having botanical allusions. The city was a typical Roman city-fort with temples, baths, aqueducts, and an efficient sewage disposal system. There was a central marketplace, now the Piazza Vittorio Emanuele, and the city was walled. Apparently there was some attempt to drain the swampy area then outside the city walls. Although flooding in winter did not assume sufficient importance to record, the Romans were well aware that snow melt was a danger, for they did erect levees on the banks of the now-contained, but still free-flowing, Arno.

The ancients knew the value of woodlands as flood protection. Plato deplored the reckless cutting of the Greek forests and pointed out the cause and effect relationship between forest destruction and the rapid runoff of rain from bared hills. Pliny quoted Plato when he admonished the Romans to maintain and to replenish their forests, and while we have no direct evidence, it seems likely that some form of forest conservation was practiced in the Tuscan hills.

With the population increasing, terracing also increased, but this stopped rather abruptly. Malaria, which had been introduced into Sicily from North Africa during the fourth century B.C., moved onto the continent during the Second Punic War of 218-204 B.C. By the beginning of the Christian Era, it reached Tuscany, where it became endemic in the lower valleys and in the city. Within less than a century, Florence and the hill towns had virtually been abandoned. The terraced farms, no longer cultivated, were subjected to erosion that washed away the precious soil. Lacking stability, it is easy to visualize the stripping of the soil from the hills, the mud slides, and the consequent silting of the tributary rivers and, eventually, of the Arno itself. Although no records exist, we are fairly safe in saying that minor flooding occurred in the Arno basin from about the middle of the first century A.D.

During the subsequent 300 years, the town of Florence continued to decline, local trade decreased, and the once relatively important Roman city became little more than a village. As the Roman Empire began to fall apart during the third and fourth centuries, tilling declined and the plowed land and terraced lower hillsides were subjected to direct erosion. The upper hills, however, began to restore themselves, and flooding probably decreased.

The fortune of the city took a turn for the better by the end of the fourth century as the malaria began to disappear for reasons that are not known. The Lombards made Florence into a military center, and when the Franks replaced the Lombards in 774, the entire region assumed an increased role as a trade center. By then Fiesole had declined to the level of a town, a condition that has never been reversed.

As the population grew, there was a greater demand for wood to be used for construction and for fuel. The hills were rapidly stripped, so that by the end of the seventh century wood had become so scarce and expensive that stone became the primary construction material. From the perspective of the art historian, this had some importance. The artisans of the time were both good craftsmen and good architects. These early stone buildings served as models for the Renaissance master builders, who made Florence a place of enduring beauty and, incidentally, laid out a city especially suited for flooding. As the city began to assume its present shape virtual bathtubs were created there—piazzas and squares with a single inlet street and one or a few outlets. Water rushing into one of the squares could not easily pour out the other side and the whole area filled up. Most of the listed high-water marks of the 1966 and earlier floods occurred in these piazzas.

Although the Etruscans and the Romans engaged in the wool trade, and Florence had been a weaving center for many centuries, the increased use of wool cloth by peoples less exalted than the aristocracy resulted in a great increase in woolen mill activity. As the woodlands were decimated, the cleared land that resulted became pasture for goats and sheep. Now, both these animals nibble forage to the root level, effectively killing the individual plant. The overgrazing became so severe by the eighth century the Florentine mills could no longer depend on local wool and hair, but were forced to import it from England and Spain. Parenthetically, it is likely that the denuded Spanish hills are partially the result of overgrazing by sheep to supply wool to Florence.

The effect of land clearing on the climate of Florence is not easy to document. Modern ecological and ecophysical studies show that forested areas serve as natural air conditioners. Heat is collected by trees during the day and released slowly at night. Modified by the cooling power of evaporation, the heat never builds up as it does when captured by stone and paving; a walk from a city street into a park at dusk illustrates this dramatically. The presence of nothing but baked clay on the hills above Florence served to perpetuate its summer heat and drought. The forests served, as the Etruscans and Romans knew, to ameliorate the heat of the summer. Not incidentally, the absence of the forests with their water-holding capacity resulted in loss of scarce summer moisture and the rivers and brooks ceased flowing in the summer.

By the beginning of the ninth century Florence had about 30,000 people, and the old walls had been breached and built farther out several times. The region had many different rulers, none particularly firm or effective. The weak rulers gradually let power slip from their hands, and authority, power, and ownership of the land became divided among the Roman Catholic bishops and a few wealthy landowners. These interlocked concentrations of land ownership led steadily toward the development of a feudal system in which relatively large areas were tilled by serfs for the production of a few economically important crops and products: Goats and sheep continued to graze, and meadowlands were cut from the forests farther up the hills. With the removal of the sparse grass, the summer droughts baked and caked the clay soil to the consistency of rock, and natural seeding in it became virtually impossible. Bound to squeeze the last florin of profit from the country, the church and the landowners made no attempt to replenish their land. Tuscany, a land noted for good water, forests, and fine agricultural bottomland inexorably declined into the arid, bare, and impoverished country that we see today. The greed of the church, the avarice of the aristocracy, and the ignorance of the people have given us this legacy.

The remainder of this sad tale can be told rather quickly. The first complete record of a major flood was provided by Giovanni Villani for the flood of November, 1333. There was an average of four feet of water, the city walls collapsed (never to be rebuilt), and three of the then four bridges over the Arno were destroyed; the Ponte Vecchio stood firm. Quoting Villani: "Wherefore everyone was filled with great fear and all the church bells throughout the City were rung continuously as an invocation to heaven that the water rise no farther. And in the houses, they beat the kettles and brass basins raising loud cries to God of *'misericordia, misericordia,'* the while those in peril fled from roof to roof and house to house on improvised bridges. And so great was the human din and tumult that it almost drowned out the crash of the thunder." Not unexpectedly, the flooding was worse in the low-lying districts nearest the Arno. Three hundred people died in the hard-hit districts of San Piero Scheraggio and the Porta San Piero. On the Via de Neri the flood crest was 13 feet, 10 inches, only 14 inches lower than during the 1966 flood. Villani carefully listed the losses of grain and casks of wine, and in quite modern fashion he noted the cost of bridge repair, as well as municipal outlays for cleaning the streets and for minor compensations to citizens.

Villani noted the extraordinary rainfall over the Arno watershed, but then wrote, "On May 15 there was an eclipse of the moon in the sign of Taurus . . . and then at the beginning of July there followed a conjunction of Saturn with Mars at the end of the sign of the Virgin." The church had the last word, stating that "by means of the laws of nature, God pronounced judgment on us for our outrageous sins." There was slight consolation from the King of Naples, who told the Florentines that "whom God loveth, He chasteneth." The Florentine Renaissance has opened with a bang.

A tiny note of biological intelligence was, however, struck by Gianbattista Vica del Cilento, who recommended a government-sponsored reforestation plan. Two hundred years later—the plan never having gotten off the ground—Guistino Fortunato dismissed Vico's plan, stating that trees were ugly and useless.

We need not detail the economic and biotic history of Florence beyond 1333. In spite of accurate information on causes of floods and adequate technical information on flood control, nothing was done. Vico was probably not the first, and he certainly wasn't the last, to chastise the Florentines for their laxity. Giovanni Batista Adriana said that the severe flood of 1547 badly disturbed the people who demanded that something be done. They were told that God wanted, with this act, to signify that worse calamities would occur, so they decided to wait. In 1545, Michelangelo cautioned a nephew about buying a house in the Santa Croce district, "every year there the cellars flood." His warning was apt, for the flood of 1547 was followed by another in mid-September of 1557, when the entire Santa Croce district was under several feet of water. Lorenzo di Medici warned the population about the Arno: "Its arrogant anger splits and beats away at the weak banks." But in spite of his authority to order new construction, he did nothing, and there is no evidence that the people heeded the warning.

Leonardo reported that the streams of Tuscany were muddier when they passed through populated districts and suggested that the mud was washed-out soil that clogged the streams and leached the soil of the hills. He conducted surveys and designed projects to develop water impoundments in the hills, to dredge the tributaries of the Arno and the Sieve, and to develop chambers under the city to hold excess floodwaters. These, too, were ignored.

The Tuscan flood situation had only one useful effect. In 1861, President Lincoln appointed the Vermonter George Perkins Marsh as first American minister to the kingdom of Italy. Marsh's travels in Tuscany and his reports on the relatively minor floods of 1861 and 1863 undoubtedly contributed to his thinking in the preparation of *Man and Nature or Physical Geography as Modified by Human Action,* published in 1864. As a result of the interest (indeed the controversy) created by this book, Marsh was asked to help compile the irrigation laws of Italy, laws never adequately enforced, which could have alleviated the floods that have occurred during the past 75 years. It must be noted that Marsh's book so stimulated Gifford Pinchot that he persuaded President Theodore Roosevelt to set into motion the legislation leading to the conservation polices of the United States.

For what it is worth, the statistical picture is gloomy. Since 1333, there has been a moderate flood in Florence every 24 years, a major flood every 26 years, and a massive flood every 100 years. Of the massive floods, those of November, 1333; November, 1844; and November, 1966, have been the worst. Since 1500 there have been more than 50 floods qualifying as moderate and 54 qualifying as major.

In 1797, a French engineer, Fabre, announced that the alpine torrents that flooded the lowlands were a direct result of deforestation of the heights. One hundred years later, in 1890, the French government undertook the job of reforestation in the Hautes-Alpes. Engineer Surell reported that as reforestation continued, floods decreased in number and severity. This report was widely distributed in Europe, was acted upon in Germany and Switzerland, but was ignored in Italy.

Apparently no one in Florence could believe that water could cover their city in the twentieth century. But they, themselves, place marble markers on their homes and churches "commemorating" previous floods. They, themselves, proudly indicate the high-water marks of floods dating back 500 years. They, themselves, speak of their dry, bare hills, which shimmer in the summer sun, and they cannot fail to see that their "ribbon of shining silver," their Arno, is a muddy, silted-in, malodorous stream that almost stops flowing during the summer.

The New York Times of January 2, 1969, reported that during the week preceding Christmas of 1968—two years after the most damaging flood in Florentine history—the Arno crested toward the flood mark. "Only after the river began to subside at 9:00 P.M. did the crowds move homeward." The bed of the Arno is still filled with the debris of the

1966 flood and has never been dredged. It is now estimated that the silt from 1966 alone is six feet thick.

There are two dams on the Arno, the Levane dam at Livorno, 35 miles upstream from the city, and the Penna dam, yet higher into the hills. These are power dams and do not provide for flood control. No money has been (or apparently ever was) appropriated for reforestation as recommended in 1334 by Vico, for the construction of water impoundments and for dredging the Arno as recommended by Leonardo, or for caring for the land as recommended by Plato.

It is beyond the scope of this article to discuss the storage of irreplaceable books and art treasures in the basements of Santa Croce, noted by Michelangelo as the district where the cellars flood every year. It is beyond the scope of this article to discuss the reasons for the lack of coordination between the two dam sites. It is beyond the scope of this article to discuss the lack of retaining walls bordering the Arno; even Lorenzo did nothing about them. As George Santayana warned in his *Reason in Common Sense,* "when experience is not retained . . . infancy is perpetual. Those who cannot remember the past are condemned to repeat it."

14

"The King Besieged"

by Jack Hope

Jack Hope's article concerning Mineral King National Game Refuge is the product of exhaustive research, which took him from Washington D.C., to Los Angeles, from Disneyland to the High Sierras.

Standing upon one of the treeless peaks that surround California's Mineral King Valley, surveying the wilderness of the southern Sierra Nevada, provokes a certain spirit of uneasiness. The feeling is not unpleasant, but is disturbing only in the sense that it is one that we aren't used to dealing with.

I don't think it's the height that brings about the sensation, or the vast abyss that suddenly opens in front of you. Nor is it simply a product of the powerful alpine panorama. More than anything else, it seems to stem from a feeling of inappropriateness on the onlooker's part. But then—perched atop a mass of granite that thrust its way from the earth's crust 30 million years ago; gazing down from a height of 12,500 feet upon an incredible array of jumbled snow-streaked peaks, alpine meadows, and evergreen-covered mountainsides; recalling a passage from the guidebook, which notes that 200 million years earlier this land rested at the bottom of a sea—what can you do that "fits" the occasion? Take a picture? Identify wildflowers?

I don't mean to imply that Mineral King (commonly known as Mineral King National Game Refuge) is unique. In fact, most of the Sierra Nevada contains similar features—the mountains and the giant sequoias, the glacial lakes, and the wide eyed mule deer. However, as wilderness (with the exception of a few hundred acres) it is at least exceedingly rare. For in the United States only about 2 per cent of our land falls into this "wild" category—about one-fourth the area now taken up by cities.

One of your first reactions to Mineral King is that you would like to show it to others who have not had the chance to see it. Your friends primarily, but others as well: a delegation of lawmakers, perhaps, or the boys who drive grocery carts around Manhattan. It seems that if more people could see such places, the job of caring for them would be much simplified—people would want to protect them for themselves, for their children, and for their friends.

There is some truth to this thought, but there is a great naïvete as well. For wilderness is by definition a perishable commodity. It can easily be loved to death in the attempt to display it to the greatest number of people. Building roads, for instance, to "make wilderness more accessible to the millions"—despite its democratic overtones—is self-defeating; something like carving up Michelangelo's statue of "David," so that each of us might own one splinter.

For another thing, not everyone looks at the land in the same way. Mineral King's 15,000 acres are regarded by different people as a setting for a beautiful photograph, a hike, a fishing trip, or as a place for the study of biology. Some see the terrain as ideally suited for a ski resort, while others welcome the wild country as a means of getting out of contact with other people for a few hours or a few days. There are those who classify the land and its creatures as "natural resources" to be used for the generation of income; and some who regard it as important in itself—whether or not it is seen, smelled, heard, touched, or tasted by humans—its own reason for being.

Each group has its own notion of what should, or should not, be done with the land, and as you might guess, these notions conflict at least as often as they coincide. Yet, within the past 20 years, each faction has come to regard itself as the champion of a cause called "conservation" (whether their real interest lies in hiking, in hunting, or in cutting trees). This is the situation at Mineral King, where, within the next few months, the proponents of one of these viewpoints will be awarded the privilege of shaping the future of this region according to their own conception of this popular term.

On the surface, it is not always clear whether a group's choice of the conservationist label is appropriate—whether it stems from a misuse of the dictionary, or whether it is, in fact, a product of a good public relations department, which knows that the word has a high "acceptance value." The story of Mineral King may be instructive in this regard; may serve as a prism through which the apparently single beam of "conservation" can be separated into its components for the purpose of closer examination.

In February, 1965, the Department of Agriculture's Forest Service issued a *Prospectus for a Proposed Recreational Development at Mineral*

King in the Sequoia National Forest. The purpose of this rather stark little document was to solicit bids from private developers for the privilege of constructing and operating a year-round recreation resort within the Mineral King National Game Refuge, a 15,000 acre neck of Forest Service land protruding into the southern boundary of Sequoia National Park. In addition to serving as a wildlife refuge, the region is now used for a variety of activities—camping, hiking, fishing, and sightseeing.

This is not a new type of activity for the Forest Service, whose guiding principle of land management is that of "multiple use." What this term means in regard to the natural environment is that the Forest Service is sanctioned to be one degree less protective than is an agency such as the National Park Service. Thus, on the 187 million acres of public land administered by the Forest Service, a wide variety of activities are permitted, including lumbering, hunting, livestock grazing, mining—even some private cabin building.

And within the last decade the focus of national attention upon the country's recreational potential, combined with the spiraling magnitude of the billion dollar recreation business, has "sensitized" the federal government's largest landholding bureaus to the task of accommodating a prodigious number of special recreation interests, ranging from motorboating to motorcycling and from skiing to skimobiling. Toward this end, the Forest Service has undertaken to make much of its domain available for a multitude of uses, and in the case of skiing, has leased public lands to commerical developers who now operate such well-known ski meccas as Aspen, Vail, and Sun Valley in the west; and Mt. Snow and Sugarbush in the east.

Thirty-seven of the 190-some ski areas on Forest Service land are located in California—more than in any other state. Nevertheless, that state's current population of about 19 million is growing rapidly; it is expected to reach 25 million in 1976, and perhaps 40 million by the year 2000. The use of the state's outdoor recreation facilities—both winter and summer—is undergoing an even more rapid expansion. The Forest Service predicts that use of California's national forests for winter sports alone will climb from the present level of about 2 million annual visits to 6 million in 1976 and 12.4 million in 2000.

An interesting aspect of the country's recreation "boom" is the extent to which it is a middle-class (or at least a "non-lower" class) phenomenon. Participation in even the simpler forms of outdoor recreation calls for a certain income level. Gaining access to a national park or forest, for example, is frequently contingent upon automobile ownership. And such pursuits as skiing, motorboating, and the more elaborate forms of camping require a considerable investment to purchase and operate the necessary equipment, and perhaps to pay for the privilege of using it.

The typical ski-family in the western states, for instance, has an annual income of $9,500 ($2,000 above the average income for that area); an adult member of this family owns about $300 worth of ski equipment, and spends over $20 per day on his ski trips. Sporting-goods salesmen, restaurant owners, and motel operators are fond of these people.

In addition to their relative affluence, ski enthusiasts are generally well organized, vocal, and therefore influential, even though they make up only 1 to 2 per cent of the total population. And, like many special-interest groups, skiers have regular means of broadcasting their requests that new regions be opened to their sport—the many ski magazines, the political friends, and so on.

Influences of this sort play a part in the decisions made on what to do with the country's public land. Some twenty years ago, for example, the Forest Service decided to develop Mineral King Valley, hoping perhaps to relieve the pressure skiers were exerting to have other areas opened, such as the San Gorgonio wilderness in southern California.

But in 1949, when the first attempts were made to attract private capital to develop Mineral King, no bids were received, because the only access to the area was, and still is, a serpentine, twenty-five mile road leading from the San Joaquin Valley up into the Sierra Nevada range—somewhat difficult to drive during summer months and completely blocked with snow during the winter. Potential developers were not interested in building a ski resort that no one could get to. No sizable recreation complex could be installed without widening and straightening the roadway.

In addition, the road winds across eleven miles of Sequoia National Park before dead-ending within Sequoia National Forest at the small cluster of summer cabins and public campsites that make up Mineral King village. Thus, developers may have been intimidated by the legal and political complications of interfering with a road crossing land controlled by the National Park Service.

At the time the 1965 prospectus was issued, no solution to the access problem had been found; but in the 16 years that had elapsed since the first prospectus, knowledge had spread of the Forest Service's interest in the area. Learning of the intentions to develop the area's ski potential, ski clubs, developers, businessmen, and politicians in Tulare County (where Mineral King is located) began to regard the resort, along with its access road, as a real and desirable possibility. One

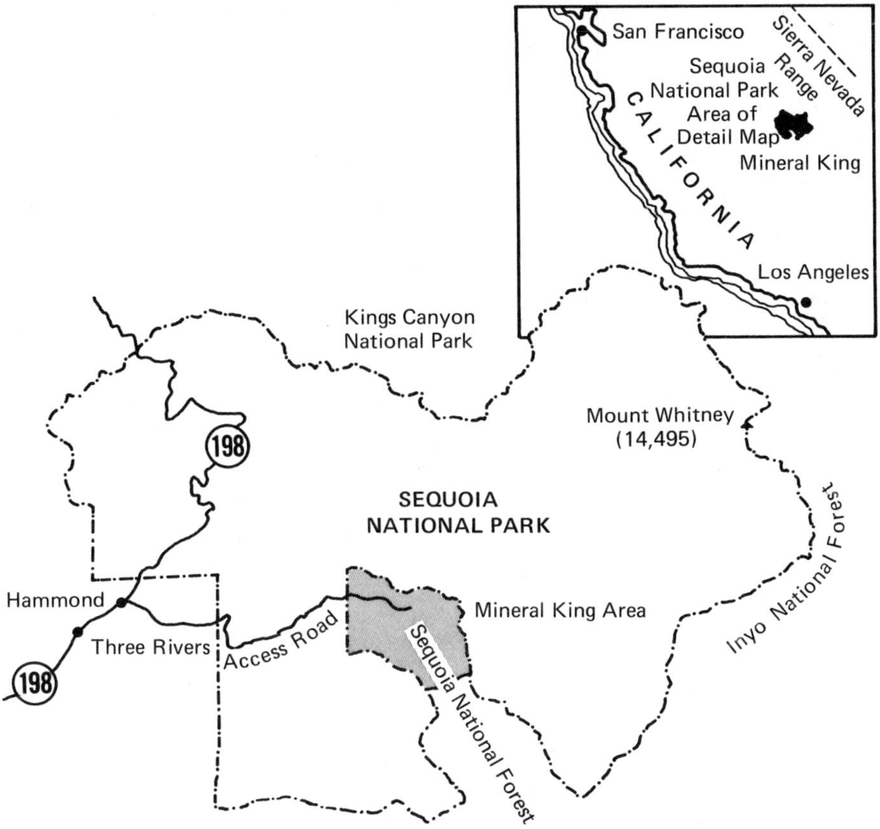

FIGURE 14-1: Maps show location and detail of Mineral King National Game Refuge. The proposed new highway (276) would closely follow the present access road.

hopeful developer (Walt Disney Productions) had speculatively purchased a small amount of private land in Mineral King Valley. Furthermore, the rapid increase in the number of California skiers had expanded the undertaking's potential support. Thus, in issuing the 1965 proposal, the Forest Service had some cause for believing that public and political support could be mustered for both the project and its necessary road.

The optimism was well founded. During the summer of 1965, the California state legislature voted to add a new highway to the state's road system. Highway 276, from the town of Hammond to Mineral King, would replace the old access road, would be kept snow-free by the state (the old access road was county maintained), and would ostensibly be wide enough to accommodate sufficient traffic at the proposed resort. The road is planned as a relatively straight, two- to three-lane highway, capable of handling about 1,200 automobiles per hour at speeds of about 50 mph. In addition, six bids were submitted to the Department of Agriculture for construction and operation of the recreation complex. These were narrowed to two, and finally, late in the year, a winning bid was announced. In December, Walt Disney Productions was chosen to carry out the Mineral King development and was awarded a 3-year permit to perform its initial studies of the region.

Up to this point, little opposition had been expressed to the project, except by local groups of outdoorsmen and by cabin owners in the Mineral King area who would be displaced by the development. National organizations that might normally oppose the area's commercial development were preoccupied with the problems of dam building in the Grand Canyon, cutting of redwoods (in the proposed Redwoods National Park), and other well-known issues. Some feared that opposition to the Mineral King ski resort would only serve to increase pressure by skiers to develop the San Gorgonio wilderness and were therefore reluctant to take a stand. Others, thinking outright opposition too late, felt that the best strategy was to try to limit the project's size. A few were even unsure of Mineral King's whereabouts, let alone its significance. But when the magnitude of the Disney development proposal was disclosed, the traditional champions of the natural environment—the Sierra Club, Wilderness Society, and National Parks Association—quickly realized that they were confronted with another major battle.

The Forest Service's original guidelines for the recreation complex were relatively modest: three or four ski lifts, parking for 1,200 automobiles, and resort accommodations for 100 overnight visitors—tempered by the warning that all facilities would have to remain "compatible with the aesthetic values of the area." The Disney firm, however, is known for doing things in a big way, and Mineral King is no exception. The original Disney plans called for a $35 million investment to include an alpine village, a heliport, a golf course, a reservoir-skating pond, swimming pools, an auto service station, two hotels, a 500-room dormitory, 1,200 cabins, a 20-acre parking lot for 2,500 autos, and ten restaurants within Mineral King Valley and on surrounding mountain peaks. In summer general sightseeing would be the main attraction; in winter skiing would bring as many as 20,000 visitors into the valley each day, from whence they would be whisked into the high country of the Sierra Nevada wilderness aboard 20 overhead lifts, which would radiate from the valley. Some of the lifts' towers would go beyond the boundary of Forest Service land and would be anchored in adjoining Sequoia National Park. Supplemental plans revealed the firm's interest in a facility, located well within Sequoia Park's Hockett wilderness area, that could accommodate cross-country skiers. Walt Disney Productions expects an annual income of about $23 million from the resort during the first years of operation, with this sum increasing thereafter as the area's popularity grows. The Disney plans also noted that "Mineral King's great natural beauty must be preserved at all costs."

With the plans before them, opponents of the development felt pressed by a twofold threat. First, they feared the immediate damage to the alpine environment that would result from construction of a roadway and a resort, and from the heavy human traffic. The prospect of bringing 2.5 million visitors annually into the heart of the mountain country conjures up pictures of tourists picking the ground clean of unusual rock fragments, wildflowers, evergreen cones, and other portions of the alpine landscape; of skiers watching the white wrappers of their candy bars slowly floating to the ground after being tossed from the gondola of one of the aerial tramways; of erosion caused by construction activity; sewage pollution in the Kaweah River; and of the intrusion of automobiles, buildings, and large numbers of visitors upon the mountain landscape. In addition, project opponents object to the disregard by the Forest Service and Walt Disney Productions for Mineral King's current status as a national game refuge. Among naturalists aware of the development plan a favorite parlor game during the past few months has been the casting of Smokey, the Forest Service bear, and Disney's Mickey Mouse in Simon Legree roles—dispossessing their fellow creatures in order to build the Mineral King resort ("Smokey strikes again!").

The Forest Service is well aware of the potential danger to the Sierran landscape posed by the planned resort. They note that "Mineral King can 'stand' development to only a certain level," but also contend that "Walt Disney Productions does not plan, and the Forest Service would not permit, a development beyond that capacity threshold." But "capacity threshold" lends itself to subjective interpretation, and what the term means to the Forest Service is apparently quite different from the meaning given by most of those opposed to the project. To the former group, it probably means a general attempt to keep ecological damage to a minimum, whatever the level of construction and human visitation. The latter define the phrase in terms of wilderness, in which case the threshold will likely be exceeded on the resort's first day of operation.

In a way, these differences of opinion reflect the essential conflict between the protectionist groups such as the Sierra Club and those more inclined toward development. Both are aware of the perplexing problem of providing recreational opportunities for a population that doubles every forty years, while our static land supply stubbornly refuses to expand despite the demands placed upon it. However, their approaches to this situation are quite different. The development faction seems to feel that the way to alleviate this difficulty is to make increasing portions of the environment easily accessible to the current demands of the population so as to spread the "visitor impact" over a larger land area. (Walt Disney Productions, for instance, notes that "the value of much public land . . . is lost because the land is inaccessible.") On the other hand, the protectionists view the growing number of claims upon the land as reason to stiffen its protection, perhaps to the point of rationing its use. This "pinch," they hope, will forcefully focus the attention of hunters, highway engineers, wilderness lovers, and resort builders on the real source of the growing real estate shortage, and will elicit stronger public sentiment to deal with the problem before the remaining public land is consumed.

From the viewpoint of the Sierra Club and its allies, the greatest threat posed by the Disney recreation complex and its access road is the influence these developments would exert upon the future protection of wild lands of the southern Sierra Nevada surrounding Mineral King Valley. Most of this land—in both Sequoia National Forest and Sequoia National Park—meets the criteria for "formal" wilderness status as defined by the 1964 Wilderness Act, and is now being considered for legal protection according to the provisions of that act. However, the presence of a high-grade highway bringing millions of people into the heart of the region would probably diminish its wild character in short order, obviate the inclusion of this land within the wilderness system, and encourage less rigid protection of adjoining land. In addition, there is some fear that the presence of a high-standard road reaching halfway across the Sierra Nevada range will be used as a rationalization for extending the pavement another 30 miles to create a new trans-Sierra highway through one of the country's last enclaves of primitive landscape.

In a way, it is tempting to apply the philosophy of "conservation democracy," advocated by Gifford Pinchot and Theodore Roosevelt, to the Mineral King situation: "the purpose of Conservation is the greatest good of the greatest number for the longest time." This concept invariably underlies the arguments in nine out of ten land-use conflicts, but despite its seeming simplicity, it poses questions that cannot be readily answered. Who can say, for instance, whether a ski resort, a semideveloped campsite, or a primitive landscape will be of greatest benefit to people who have not yet been born?

Nevertheless, the concept stimulates some interesting speculations about the future. In the case of Mineral King, you might ask what needs will characterize Californians in 2068: The need for convenience or for diversity? For material comfort or challenge? For sociality or solitude? Which need will be most in demand and least in supply?

Even at present, there seems to be a considerable difference of opinion over the proper application of these democratic guidelines. Both the Forest Service and Walt Disney Productions maintain that a new high-standard roadway into the Mineral King country will represent an improvement over the current situation because it will permit the rapid influx of large numbers of visitors ("Many people who wish to see and enjoy Mineral King but won't drive the present difficult road."). To this argument however, their opponents reply that the area's present function of providing general recreation is inherently more democratic, for it does not impose an admission fee upon visitors. They further contend that in terms of meeting California's most urgent recreational needs, the $35 million to be spent on the Mineral King project might be put to better use by expanding the recreational facilities available to the non-mobile population in, say, the Watts ghetto of Los Angeles.

Throughout 1966 and into early 1967, the Mineral King issue continued to be fought at a local level. In that time, proponents of the Disney project succeeded in gaining political support for the venture. In September of 1966, Walt Disney and California's governor, Edmund G. (Pat) Brown, flew into Mineral King by helicopter to survey the area. Thereafter, Brown made a short speech in which he described the proposed development in glowing terms: "I hope that ten years from today I can stand here with Walt Disney again and look around at the wonderland that will have been created," and assured his audience that "we are going ahead with the [access] road." The Brown administration applied for and received a $3 million federal grant to partially finance the road's construction; the California Highway Commission then voted to provide the remainder of the needed construction money (about $20 million) from the state's highway tax fund.

These actions did not pass unchallenged. California's State Highway Engineer, J. C. Womack, attacked the financing of the Mineral King road stating that the highway could proceed only "at the expense of other critical [road building] projects," and that the use of funds set aside for other construction would prove "very disruptive to previously approved planning and scheduling of projects in the Southern Counties group." State Assemblyman Alan Sieroty assailed the proposed destruction of natural landscape, stating that the plans for the Mineral King road were "atrocious," then added: "It would be . . . outrageous for the Reagan administration [Reagan was elected in 1966] to give away more than $20 million to facilitate a private venture when it has just cut the state Parks and Recreation budget . . . from $33 million to $1 million." In addition to this objection, two bills were introduced into the California state legislature in an attempt to block construction of the access road. Both died in committee before being brought to a vote.

By early 1967, although funding had been approved for the access road and its proposed routing established, and although newspapers in Tulare County were speaking as if the Disney resort was already a fact, the National Park Service had not yet granted permission for the road to cross eleven miles of Sequoia National Park on its way to Mineral King Valley. It is still unclear whether the Department of the Interior can legally grant permission for a road to cross a national park for non-park purposes, or whether this requires an act of Congress.

In any event, Secretary of the Interior Stewart Udall disapproved of the new road. He noted that it would prove a "blighting influence" upon Sequoia Park, that blasting of rock and planned removal of eight million cubic yards of soil would scar park terrain, cause erosion and consequent siltation of the Kaweah River, pollute the air with auto exhaust fumes, pose a threat to giant sequoias and other large evergreens, and would, in general, compromise "park values." With a first-hand sensitivity to the proliferation of the nation's vast network of highways (3.6 million miles worth) he commented that developers "will be more honored 25 years from now for the roads they do not build than for the roads they do," and suggested that the Forest Service and Disney Productions explore other means (such as an overhead monorail or tramway) of transporting sightseers and skiers into Mineral King Valley to minimize destruction to the landscape.

To answer charges of road damage, the California Division of Highways commissioned biologist Richard J. Hartesveldt, an authority on giant sequoias, to study the ecology of sequoia groves surrounding the proposed roadway. The report's objectivity, however, has provided at least as much ammunition for opponents of the project as for its advocates, and each side has quoted and paraphrased the document upon numerous occasions.

As for Udall's proposal that a monorail or tramway be used to transport visitors into the resort, this plan was rejected by both Disney Productions and the Forest Service as being too expensive. The cost of installing such a system—unlike the tax-financed roadway—would be borne by the Disney firm.

By mid-year 1967, opponents of the development had finally managed to air some of their grievances to a national audience by means of articles printed in magazines published by the National Parks Association and the Sierra Club. Within these groups, the general feeling seems to be that the only way to provide Mineral King with adequate protection against commercial development is to make it a part of adjoining Sequoia National Park. In 1926, when Congress was wrestling with the problem of what to do with this area, the decision was made to award it to the Department of Agriculture in the belief that there were minerals remaining in and around Mineral King Valley that would become "locked up" if park status were granted. But since no mineral deposits have been discovered, there is a strong feeling now that these 15,000 acres should be given to the Department of the Interior.

This type of interdepartmental land transfer is invariably an unpopular procedure on the part of the donor agency, and is avoided by Congress whenever possible. Just now, the Forest Service is especially sensitive on this count for portions of forestland are being transferred from the Forest Service to the Park Service in order to establish a Redwoods Park in California and a Cascade Mountain Park in Washington. The history of transactions of this sort carries all the flavor of a range war between sheepherders and cattlemen, and while reading these accounts, it is easy to lose sight of the fact that the ownership of federal property rests with the public rather than with the managing agency.

By this time, Secretary Udall was finding it increasingly difficult to maintain his opposition to the access road upon which the success or failure of the development depends. A hint of behind-the-scenes pressure appeared when the Secretary indicated that he would need a great deal more support from leaders of the protectionist movement than they had given him. In addition, the open opposition to Udall's position had assumed formidable proportions. California's two senators, several representatives, lieutenant governor, many state legislators, and Los Angeles Mayor Yorty now stood solidly behind the Disney plans. Senator Thomas Kuchel, who can usually be counted on to champion the position of the Wilderness Society and the Sierra Club, reversed his role in the case of Mineral King and went so far as to write an article in a ski magazine praising the merits of the Disney firm and its proposed undertaking. Real estate speculators and businessmen in

Tulare County, anxious to capitalize upon the anticipated influx of tourist gold, were beginning to make impatient noises. These were echoed by Tulare County politicians and newspaper columnists, and by Secretary of Agriculture Orville Freeman, all of whom viewed Udall's reluctance to approve the access road as an obstructionist tactic.

To add more fuel to the fire, Governor Reagan sent emissaries to persuade Secretary Udall to change his stand. Although Udall remained adamant, he was misquoted by a Reagan aide as having given his approval to the Disney project. The story that Udall had conceded, followed quickly by his denial, served only to heighten tension and to increase pressure for a decision.

Of the obstacles faced by Stewart Udall and his supporters, one of the most formidable is the intangible something which Walt Disney Productions calls the "Disney Image." For after being honored by a number of nature-oriented organizations, ranging from the National Audubon Society to the National Wildlife Federation; after thirty years of producing such films as *Bambi, Beaver Valley,* and *Nature's Half Acre,* it hardly seems possible that the Disney firm would consent to undertake a project that would in any way prove detrimental to the great out-of-doors. The common conception of the Disney orientation toward the natural world, as Disney publicity states, is one of a "proven interest in conservation." As noted in the organization's proposal for the development of Mineral King, issued to the Forest Service: "The integrity and quality image of Walt Disney Productions is *already* pre-sold in the public mind." This seems to be the case, particularly in California, where considerable fame and fortune have come to rest since the establishment of the Disney empire in that state.

In July, 1967, the nature of the suspected pressure on Secretary Udall was made public. To the consternation of those who had labored to defeat the Mineral King proposal, it was revealed that through the complicated machinery of Washington diplomacy, the fate of Mineral King had become linked to the ongoing negotiations between various federal agencies and the state of California over the size and location of the proposed Redwoods National Park. Apparently, the Johnson administration felt that Governor Reagan would be more amenable to the installation of a national park in his state if he were offered certain concessions in return. One of several concessions offered the state was the administration's active support of the Mineral King development and its access road. The nature of this transaction was spelled out in a letter of June 22, 1967, from an administrative spokesman. Phillip S. Hughes, Deputy Director, Bureau of the Budget, Executive Office of the President:

> Following is the Administration position on a number of items over and above the provisions of S. 1370 [the redwoods bill introduced by Thomas Kuchel]. . . . The Administration is prepared to implement these provisions immediately. . . .

The Mineral King proposal is listed under a section labeled "Other Conservation Program Actions."

> 6. *Mineral King*
>
> It is in the interest of the Administration and the State that the Mineral King area . . . be . . . developed. . . .
>
> The Department of the Interior has been requested to consider issuance of a permit jointly to the Department of Agriculture and the State of California for a two-lane road through the Park [Sequoia] to provide access to Mineral King. . . .

The letter's contents provoked a volley of angry protests from the project's opponents. Senator Lee Metcalf (Montana) assailed the administration's intention to "override the Secretary's [Udall's] best judgement" as raising "fundamental questions of public policy, precedent and equity." Anthony Wayne Smith, General Counsel and President of the National Parks Association stated that the Forest Service had "put to auction a priceless asset [Mineral King] belonging to the American people. . . ." California Congressman Phillip Burton perhaps best summarized sentiment against the maneuver when he angrily labeled the Mineral King-Redwoods Park trade as "a desperate effort to pay the high political price needed to get along with California's new governor" and bemoaned the fact that the executive branch "promises to pressure the National Park Service into compromising the integrity of Sequoia National Park by accommodating the opening wedge for massive development of Mineral King."

Whatever the processes by which pressure was exerted, Burton's fears were apparently justified and the intent of Hughes' letter carried out. In December, a high-level Mineral King meeting was called. Attendants in favor of the project included: Secretary of Agriculture Freeman, California Senators Kuchel and Murphy, Congressman (and former Olympic decathlon champion) Robert Mathias, and for the administration, Phillip Hughes. Attendants opposed: Stewart Udall. Shortly after the conference, Freeman issued a press release announcing that the right-of-way for a road through Sequoia Park was being prepared by the Interior Department and that the Disney development would proceed as planned.

From the point of view of the project's supporters, it probably would have looked better if the release had come from the Department of the Interior with Secretary Udall's blessing. This was not the case, however. To this day, Udall has not voiced support of the access road, leading to the conviction that this decision was predetermined by the administration. Even so, the Secretary was for a time, nobody's favorite. At a press conference in which he was criticized for "his" decision, Udall retorted that things might have been different had preservation groups spent more of their time writing to Orville Freeman than to him.

The Mineral King development is now within one step of realization. But this last step has proven time consuming. For in the year that has elapsed since Orville Freeman's announcement, the Forest Service and National Park Service—never the friendliest of "sister agencies"—have failed to reach agreement on the dimensions of the proposed access road crossing Sequoia National Park.

The delay has proven helpful to the more tenacious of the project's opponents; has given them time to begin legal investigations and to renew efforts to gather public support. During the past five months, two to three thousand signatures have been added to an antidevelopment petition circulated by the Sierra Club, network television stations have hauled their cameras into the Sierra Nevada to cover a "hike-in" staged in opposition to the project, and students from UCLA have pulled up and burned survey stakes marking the location of the proposed access road. Some of the more militant even plan to picket Disney films and to stage "lie-ins" in front of highway department bulldozers if construction begins, as scheduled, next summer.

It seems that the era of modern conservation weaponry is upon us. Whether or not this will aid the case for Mineral King will depend on public opinion and on the orientation of the new President-elect toward the issue. Resolution of the conflict will not be long in coming.

15

"The Ecology of a Reservoir"

by Paul A. Erickson and John T. Reynolds

Paul A. Erickson has combined a career in biology with one in sociology, which he defines as "only human ecology." He finds that each field naturally leads to the other as he moves between microbial and human relationships. Mr. Erickson is a research associate on the Water Resources Research Council of the University of Massachusetts, project manager of Clark University Biology Department, and a lecturer in sociology and anthropology at Clark University Evening College. His future studies include the sociopolitical aspect of water diversion.

After two years of postdoctoral work in marine microbiology, John T. Reynolds began his study of the Quabbin Reservoir in 1965. He is specifically interested in the ecological significance of cell-destroying bacteria in aquatic environments. Dr. Reynolds, Professor of Microbiology at Clark University, Adjunct Associate Professor at the University of Massachusetts, and senior consultant in microbiology for the New England Research and Development Corporation, took his degrees in biology, bacteriology, and microbiology.

When most people think of bacteria, they think of diseases. They might therefore consider that Albert Schweitzer was carrying "reverence for life" too far when this philosophic medical missionary applied his tolerance to the bacterial world. But one need not be a philosopher to have high regard for these ancient wonders of the microcosm. One need only be: (1) a bit thirsty and (2) inquisitive about the natural processes that provide us with clean drinking water.

Of course, to many of us a reservoir—that intricate ecological system from which we draw our precious water—exists only as "the other end of the faucet." Or perhaps it is a scenic view beyond the pines on a deserted way. In either case, it is usually thought of as a healthy reservoir only in terms of bacteria being absent. Yet, should the faucet someday run dry or the scenic view become a blight, the reason might conceivably be found in a simple fact: bacteria, necessary to the reservoir's well-being, had ceased to do their work effectively.

The problem of maintaining our reservoirs in a healthy condition is quickly assuming major proportions in our society. Microscopic plants known as algae, which can compete with some forms of bacteria for essential nutrient, are the most common cause of public concern, and their presence in large numbers is usually used by the scientist as an indicator that all is not well with a reservoir, that it is growing old and senile. Foul odors and taste in drinking water supplies, clogged filters in water treatment plants, increased costs of water treatment, grossly impaired appearance of valuable lakes, depletion of dissolved oxygen in water (and therefore the depletion of important fish populations), and greatly diminished land values—these are but a few consequences of the growth of large numbers of algae and the senility of water. They are sufficient, however, to warn us that our own physical, economic, recreational, and even esthetic lives are intricately bound up with microbiological processes that govern the life and death of our reservoirs.

In spite of our vital dependence upon these microbiological processes, it is recorded by the United States Department of Health, Education and Welfare, in a publication devoted to the significance of algae in water supplies, that "studies or attempts to control algal problems in water works have revealed hit-or-miss procedures, little coordination of effort, and only occasional systematic recording of essential data." This criticism of man's efforts to understand and control his environment was made in 1962. In 1965, one attempt at the necessary "systematic recording of essential data" on the interdependent roles played by bacteria and algae in the life processes of reservoirs was begun in Quabbin, a reservoir of special significance to the citizens of Massachusetts.

Completed in 1937, Quabbin Reservoir is approximately 60 miles west of Boston. It is a man-made, 18-mile-long reservoir, created in the Pioneer Valley area by damming up three branches of the Swift River and more than a dozen "Walden-type" ponds so characteristic of the New England landscape. The plan for Quabbin, first conceived in 1895, eventually called for the complete destruction of the towns of Dana, Enfield, and Greenwich; the partial destruction of the towns of Salem and Prescott; the displacement of 2,500 citizens; the clearing and burning of trees and shrubs over approximately 38 square miles of farm and woodland; and the reburial of more than 7,000 bodies from 34 cemeteries. The reservoir began to fill in 1939 and reached its maximum capacity of 412 billion gallons in 1946. Because of a recent drought in the northeastern states, this vital water resource fell to 50 percent of its capacity in 1966, but it has since risen to about 60 percent of capacity.

Predictions as to how long it will take for Quabbin to refill are admittedly pure guesswork. Nevertheless, thirty-one municipalities of the greater Boston area still depend upon Quabbin as their basic water supply. Several lesser reservoirs in Massachusetts and their associated water treatment plants are maintained at adequate levels only by aqueduct diversions of water emanating from Quabbin. In the summer season alone the output of clean water from this reservoir can exceed 10 billion gallons per month. Obviously the importance to man of the aging of water is directly proportionate to the importance of the water in which it takes place. Hence the significance of studies concerning eutrophication in a reservoir as important as Quabbin.

In the waters of such a reservoir there are complex, dynamic interactions of physical, chemical, and biological factors. These act over time to minimize the accumulation of free organic material in the stored water. Such organic material is always present, since much of it comes from the cellular debris of algae that live and die in the reservoir. But algae are not the only source. Other organisms and organic materials—such as leaves, commercial fertilizers, pesticides, and dead animals—are continually introduced into the reservoir by wind, stream, animals, and, of course, by man.

A significant portion of all organic material found suspended in the water or in the muck at the bottom of a reservoir, whether that material originated within the reservoir or outside of it, is cellulose. Cellulose is a complex chemical compound found in the cell walls of most plants. It is of special significance because it is not easily degraded, or transformed, into less complex compounds. Man, for example, excretes most of the cellulose that he ingests when he eats a celery stalk because his digestive system cannot break this molecule into smaller pieces. But if the cellulose in a reservoir were not degraded in some fashion, it would go on accumulating until, much like a municipal dump overflowing the city, it replaced the water.

The cellulolytic bacteria are among the few organisms present in the depths of a reservoir's water that can accomplish this necessary degradation. Meanwhile, various other bacteria decompose the remaining non-cellulose debris.

That bacteria can accomplish the decomposition of organic material

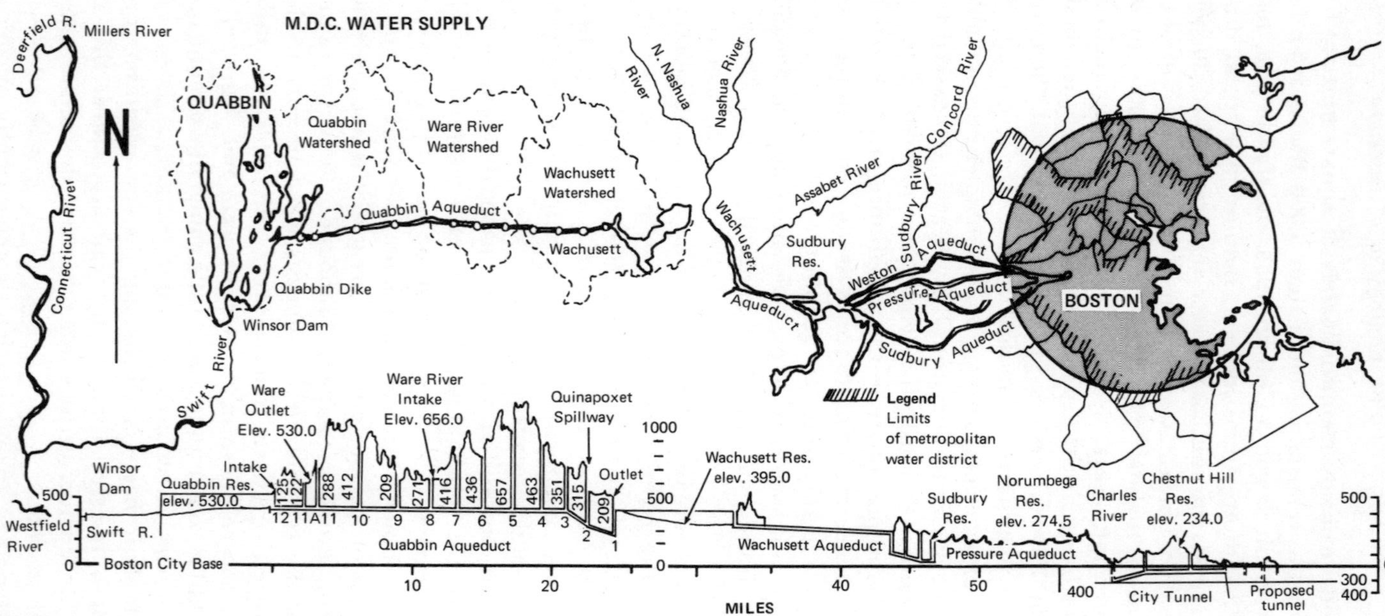
M.D.C. WATER SUPPLY
Deerfield R.
Millers River
Connecticut River
N
QUABBIN
Quabbin Watershed
Ware River Watershed
Wachusett Watershed
Quabbin Aqueduct
Wachusett
Quabbin Dike
Winsor Dam
Swift River
N. Nashua River
Nashua River
Concord River
Assabet River
Sudbury River
Wachusett Aqueduct
Sudbury Res.
Weston Aqueduct
Pressure Aqueduct
Sudbury Aqueduct
BOSTON
Legend
Limits of metropolitan water district
Westfield River
Swift R.
Winsor Dam
Quabbin Res. elev. 530.0
Intake
Ware Outlet Elev. 530.0
Ware River Intake Elev. 656.0
Quinapoxet Spillway
Outlet
125
122
288
412
209
271
416
436
657
463
351
315
209
12 11A 11 10 9 8 7 6 5 4 3 2 1
Quabbin Aqueduct
Boston City Base
1000
500
Wachusett Res. elev. 395.0
Wachusett Aqueduct
Sudbury Res.
Norumbega Res. elev. 274.5
Charles River
Chestnut Hill Res. elev. 234.0
Pressure Aqueduct
City Tunnel
Proposed tunnel
500
0
300
400
10
20
40
50
MILES

FIGURE 15-1

that settles on the reservoir bottom would seem to obviate any fears that the reservoir must quickly choke itself with debris. But, in fact, this decomposition ushers in even more immediate threats to the life of the reservoir. Just as microorganisms present within the specialized stomachs of cows, rabbits, and sheep digest the cellulose of grass and leaf, thereby providing these animals with essential nutrients, bacteria working within the bowels of a reservoir release into the overlaying waters equally significant nutrients. In turn, these nutrients are largely taken up by those very algae that increase in number by feeding on the rich chemical feed. When these microscopic organisms die and settle to the bottom of the reservoir, the decomposition cycle begins all over again. Thus, year by year, the water becomes increasingly unfit for human use.

To be sure, there are ecological situations where the uptake of nutrients by algae is proving useful to man. An example is South Africa, where water resources are severely limited. There, sewage, rich in nutrients released upon bacterial decomposition of fecal material, is inoculated with algae. These grow quickly by removing nitrogenous compounds from the sewage. The enhanced algal growth, which appears as a green slime and is commonly called "an algal bloom," therefore frees sewage water of those chemicals detrimental to man and facilitates the direct recycling of such water back into human consumption.

However, when our aim is not to purify sewage but to maintain the purity of reservoirs, the uptake of nutrients by algae—far from being a blessing—is a clear and ever present threat. Here the situation is best described in terms of a delicate balance between those forces that would solidly clog the reservoir with organic debris, and those that would increase its algal populations to undesirable levels through a bacteriological removal of that debris.

Such delicate balances are characteristic of all ecological systems and are known by the biologist as "moving equilibria." These are "moving" balances in the same sense that, while a seesaw is in "stationary" balance when two boys of equal weight are sitting on opposite ends, the balance "moves" and the board tips if a third boy jumps on at one end. In similar fashion the balance characteristic of any ecological system is continually being struck and lost and gained again for the simple reason that organisms, along with their individual nutritional requirements and products, are continually being added to, and subtracted from, the environment. When the ecological system with which we are dealing is a reservoir, we are most interested, of course, in maintaining the specific and certainly delicate balance that is most beneficial to our human needs.

The shift of this equilibrium, or balance, in a reservoir over time is referred to as the aging process of that reservoir. In terms of ponds, the "aged" (or eutrophic) pond is a body of water with such an accumulation of organic material that the bacterial degradation of that material, plus nutrients arriving from the watershed, makes a great variety of nutrients available for the support of larger populations of more complex forms of plant life. In the "young" (or oligotrophic) pond, on the other hand, the continual deposition and degradation of organic debris proceeds so slowly that large algal populations and emergent plant forms, such as water lilies and cattails, cannot exist. Every oligotrophic pond eventually becomes eutrophic, just as surely as every one of us must grow older. But unlike man, whose lifetime has been increased by modern technology, the lifetime of a pond or reservoir has been drastically shortened by that same technology. Before the advent of technological man and his power to pollute, the lifetime of a body of water could be measured in thousands of years. Today, man's demands for clean water are ever increasing. Yet he introduces into the atmosphere and water materials that serve as nutrients for algae—and an oligotrophic lake can become senile and die within a decade.

It is, regrettably, a simple fact of life that we cannot as yet stop or reverse the eutrophication process going on in our water resources. Therefore the people responsible for maintaining adequate supplies of clean drinking water, such as those at Quabbin, are primarily interested in techniques designed to slow down the aging process. A number of these techniques are based upon the following procedures, either singly or in various combinations. There are many other procedures, of course, but these eight are deemed most applicable for the long-term control of eutrophication in established reservoirs. They call for:

1. Reduction of nutrients introduced into the reservoir. This is done through strict control and surveillance of sewage waste disposal and chemical fertilizers, which could leach into waters draining into the watershed of the reservoir.

2. Reduction of nutrients released into overlaying water by bacterial decomposition of bottom sediments. The nutrients are held back by the human deposition of a non-organic, inert "covering layer" on top of those sediments.

3. Reduction of algal populations by mechanically harvesting the algal crops. (This reduces the amount of nutrients released upon their decomposition.)

4. Reduction of organic deposits on the reservoir bottom. This is done by mechanically harvesting rooted and blossoming plants.

5. Removal of organic deposits on the reservoir bottom by dredging.

6. Removal of dissolved nutrients from water by the use of physical and chemical agents.

7. Removal of algal populations by chemical agents.

8. Reduction of the algal population by stimulating an increase in the fish life that feeds upon those populations.

While each technique offers some hope that the steady progress of eutrophication might at least be temporarily thwarted, none is yet integrated into programs considered universally practicable and efficient for the continual day-to-day depression of that process in our reservoirs. One reason for this situation is that many of the techniques are still in the experimental stage. It could well happen, for example, that dredging a reservoir's bottom muck would be too drastic an operation—stirring up the muck might increase the amount of nutrients dissolved in the water. The same result might come from mechanically uprooting plants on the reservoir bottom.

Also, the various procedures have not yet been incorporated into a workable program because even if they should prove scientifically possible, the financial and physical problems associated with their actual application are so immense as to make them, for now, impracticable. For example, even if we knew precisely which nutrients to extract from the water, and even if we had the materials and techniques necessary for that extraction, could such techniques be realistically applied, say, to more than 400 billion gallons of water in a single reservoir? To develop procedures out of studies conducted on small "laboratory ponds" is one thing; to apply them to large reservoirs is another.

Algicides such as copper sulfate are often employed in the effective control of algae in water supplies. The effectiveness of these chemical agents, however, depends upon a number of factors: the density of the algal populations, the specific kinds of algae present, the alkalinity of the water, and the amount of organic material suspended in the water. Their use, therefore, requires continual chemical and microbiological monitoring of the reservoir. Such monitoring, given the availability of trained personnel and facilities, is not always possible.

Finally, and perhaps most importantly, it must be remembered that a reservoir is not just a body of clean drinking water put aside by society for the communal good. More realistically, a reservoir is also a focus of sociopolitical factors that must be taken into account when designing any procedure aimed at the control of eutrophication. One has only to think about the public fury generated on both sides of the fluoridation question, for example, to understand a fundamental point about water resources and methods (both current and proposed) for their preservation: the deeper the reservoir, the more complex and strong are its sociopolitical currents. This is not so much an idealistic criticism, as it is a mere observation that the ecological system of a reservoir includes our own sociology. If it is scientifically possible to dredge a reservoir, for example, is it sociologically wise to "dirty up" a fisherman's favorite lake? If it is scientifically feasible to reduce nutrient levels in a reservoir by strict surveillance of sewage disposal and fertilizer use in areas surrounding a reservoir, is it politically possible to do so? These are obviously questions that the biologist and the engineer cannot answer—but it is often within the framework of just this kind of question that their proposals must be built.

It is because of such complexities that one other possibility for slowing down eutrophication in our reservoirs is of particular interest today. It concerns the unique studies, classified under the general heading of "microbial intervention," and initiated under the direction of the Water Resources Research Council of the University of Massachusetts, that began at Quabbin in 1966 and are still in progress.

The rationale behind such studies can be simply stated: if the ecosystem of the reservoir could be so managed as to favor those groups of bacteria that already compete with algae for the nutrients derived from decomposed organic sediments, the algal food chain would be broken. Therefore algal populations of undesirable proportions could not occur. Such microbial intervention is significant because it would maximize the reservoir's self-cleansing capability—a capability that seems to have been quite sufficient up until the time of "Homo pollutans." If it could be sufficiently maximized for today's needs, microbial intervention would tend to soothe the political and social pulse. For it is, after all, a "natural process."

What is needed, however, before we can construct and implement any procedure founded upon the rationale of microbial intervention—or any other procedure for checking the eutrophication process—is precise information describing the interactions of algal and bacterial populations over time in our reservoirs. We have already learned, for example, that Quabbin Reservoir may be subdivided into several areas, each with a unique microbial ecology. To illustrate, the number of algae in water samples from one area of Quabbin is such that the water has a distinct greenish-brown color, while the algal population in another area is often only one-tenth as great, and the water is crystal clear. And we know that the kinds of algae indigenous to one area are not necessarily the same kinds present in other areas.

But we also know that when algal populations increase in Quabbin, bacterial populations generally decrease. This brings up the important question. "Why?" It remains unanswered, as yet, but only through a continual collection of this kind of information are we able to ask such questions at all—questions that pertain directly to the problem of eutrophication and that are therefore central to the future formulation of methods that will effectively deal with the problem.

In the period of time Quabbin has been under study, hundreds of gallons of water have been collected, filtered, and tested, and their contents have been examined under the microscope. In the same period of time, billions of gallons of that same water have been poured into millions of drinking glasses in hundreds of thousands of homes in the heart of Massachusetts.

That water comes into Quabbin undrinkable. It stays there, according to some estimates, for up to three years, and finally leaves Quabbin crystal clear. Exactly how this feat is accomplished remains a mystery. But more than that, as long as the mystery remains, we demonstrate that we still know too little about Quabbin's total ecology; we remain in serious danger that those same processes of self-cleansing shall become disrupted by the eutrophication process and result in the death of the reservoir.

That we have as yet no practical methods for bringing about an effective slowing down or reversal of that aging process is perhaps disappointing and definitely alarming. But it should not be surprising. We have crawled out from our caves to add cruelly to the burden of an ecological system geared to its own cleansing before we have fully understood the basic microbiological processes by which that cleansing is carried on. If we are now to keep our reservoirs alive we must inquire before we can hope to dictate: we must understand natural mechanisms precisely before we can manipulate them according to our own human requirements.

But even now, have we really turned away from those dark primordial cavern walls where we trace our own smudged effigies in a vain effort to trap the natural world in a human wish? That question will remain as long as we are preoccupied with bacteria as destroyers of human life and know so little about them as vital links in the process by which nature provides clean water.

16

"Escalante Canyon"

by Edward Abbey

The author of Desert Solitaire, *a book on the canyonlands, as well as of several novels and magazine articles, Edward Abbey received his B.A. and M.A. degrees in philosophy from the University of New Mexico. Presently a seasonal ranger for the National Park Service, Mr. Abbey's primary concerns are "the liberation of women, the preservation of wilderness, and a new and better American Revolution."*

The Escalante is a small river that flows from the high plateaus of south-central Utah into what was once the Colorado River, now the reservoir called Lake Powell. In the course of its final sixty miles the Escalante has carved a meandering, deeply entrenched canyon through the monolithic sandstone that overlays most of the region. This river is the living heart of a roadless canyon system, which is the best that now remains of the former Glen Canyon wilderness. Like the latter before its submergence under the waters of the man-made lake, the Escalante and its side canyons contain a variety of bridges and arches, seeps, springs, streams, and waterfalls, secret garden alcoves and towering slickrock walls—a peculiar combination of natural features that cannot be found anywhere else, so far as I know, but in the canyonlands of the Colorado Plateau.

One day last spring I went for a walk down into the lower end of the Escalante and its tributaries. Leaving my car at the head of Hurricane Wash, some forty miles southeast of the town of Escalante, I tramped down the wash toward the semisubterranean world of the canyons. On my back I carried a light sleeping bag, enough dried food for three days, and a gallon of water, plus such miscellaneous (but indispensable) items as matches, a notebook, and old tennis shoes.

The trail begins by following an old jeep track (a rancher's dead-end road) for a couple of miles; then the track diverges onto high ground while the trail continues in the sandy, descending bed of the wash. The way was well trodden by the hooves of many cattle—the property of local ranchers—and not a few hikers, even by the wheels, I sadly noted, of a motorbike. This is primitive country, not a pristine wilderness, and it has already begun to attract hikers and campers from all over the nation. As I saw by the scratchings on a slab of sandstone, a whole troop of Boy Scouts had passed this way only a year before. (I should mention that Hurricane Wash is the easiest and most popular entryway to the Escalante; its other side canyons are seldom entered.)

Gradually the wash deepens, becomes what in most parts of the West would be called a canyon. The buff-colored sandstone walls rise almost vertically on either side, unscalable, for the most part, except with climbing equipment. There is some plant life, but not much; the canyon floor consists of waterless sand and little else. The walls close in and for the distance of a crosstown Manhattan block I marched through a corridor of nothing but sand and sandstone: lifeless, a deep narrow channel of architectural symmetry in the rock, scoured so vigorously by occasional summer floods that no plant of any kind has been able to take root. The corridor is about eight feet wide, about thirty or forty feet deep; to be caught here during a flood would be fatal.

Then the canyon broadens as it deepens. The first signs of water appear: wet sand, mudbanks, swaths of pickleweed and wire grass, clumps of willow. And deer prints and cliff rose, cow dung and flies.

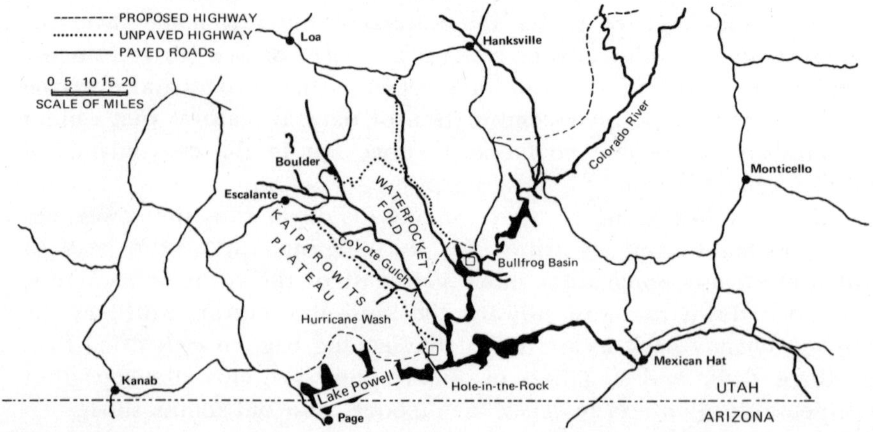

FIGURE 16-1

Dried algae on the rocks, tiny, stagnant pools alive with water skippers, dragonflies hovering above and swarms of mosquito larvae within. In a deep-shaded undercurve of the wall I came to the first seep and a perceptibly flowing rill of water. One could drink here, if necessary, although the water is polluted by the infestation of cattle. I trudged on without pausing; although hot and thirsty, I had the four quarts of water in my pack still untouched. Besides, I wanted to see how far I could go in comparative comfort without a drink. In this way I'd find out how much water I would need for the hike out, back to the car.

Deeper and deeper the bed of Hurricane Wash cuts into the bedrock; the walls become several hundred feet high, much higher than the canyon is wide. The series of pools join one another, forming a trickling flow that oozes over the rock and mud and sand. I passed a few cottonwood trees, growing tall and slender in this deep, shadowy canyon. More thickets of willow and tamarisk, many tracks of deer, cattle, and smaller beasts. From somewhere ahead, around a bend in the canyon, I could hear the alluring sound of running water. I hurried on, turned the corner, and saw that the canyon I'd been following now joined a much bigger canyon. Here were beaches of sand, groves of giant cottonwoods, and down the center of the canyon floor, issuing from a narrow gap in the walls, a clear-flowing stream several feet wide.

I removed my pack, sat down on the sand against the shady side of the wall, and checked my map. I'd come about five miles and had now reached Coyote Gulch, one of the major side canyons of the Escalante. The Escalante itself was still at least five or six miles away, as near as I could judge from the intricately meandering course of Coyote Gulch. I rewarded myself with a good drink of water, switched from boots to tennis shoes, and went on down the gulch, following the stream bed. This made pretty good walking most of the way—the water was shallow and the sand was firm.

Rock and water, sheer cliffs, and delicate details. On my right I saw a thread of water falling from a seep on the undercut wall of the canyon. I stepped over to the seep holding out my tin cup, and the water tinkled into it with the most musical of desert sounds. Overhead a garden grew, a hanging garden of panic grass, maidenhair fern, scarlet monkey flower, hellebore, poison ivy, and pale yellow columbine. Here would be a good place to tank up, fill a canteen or two, before beginning the walk out of the canyons. I was unwilling to drink from the creek with all those cattle upstream.

The canyon curves deeply to the left and right, sinuous as a snake, no more willing to follow a straight line than anything else in nature. On the inside of each meander the walls slope obtusely upward; on the outside, where undercut, they slope acutely, so that the curves parallel each other like a loosely jointed ball and socket. The overhanging walls resemble hollowed-out domes, some of them hundreds of feet high. Standing deep inside one of these alcoves you may not be able to see the sky at all; in such places, although damp from water seepage, there is almost no plant life, because sunlight cannot enter except by reflection. Walking in a canyon like this your points of orientation are reduced to two: upstream and downstream.

Perhaps a mile below the junction of Hurricane Wash and Coyote Gulch I rounded a bend and saw a window in the rock ahead: Jacob Hamblin Arch. Formed in the neck of a wall of sandstone that juts out into a loop of the canyon, this arch will someday become a natural bridge when the creek below completes the work of short-cutting its own meander.

The way up through the arch looked rough and rocky; with the pack on my back I preferred not to try it. Instead I followed the creek on its way around the gooseneck, passing beneath—or within—a great curved and overarching shell of sandstone. Imagine the Hollywood Bowl expanded to something like ten times its present size. The apex of this structure must be at least five hundred feet above the stream bed.

The last of the afternoon sunlight had long since vanished from the upper walls of Coyote Gulch. I walked through a lavender twilight, the murmur of flowing water, the croaking of toads. The willows swayed, the leaves of the cottonwoods rustled like paper as an evening breeze came gliding up the canyon. It was time to make camp, but the charm and magic of this (to me unknown) canyon was so great that it was hard to stop. Each turn in the walls seemed to hint of something new and even more delightful; exploring such a place is like the gradual discovery of the personality of a new friend, a new love.

But since I would not stop for dark, the dark stopped for me, surrounded me. On a shore of sand well away from the cattle paths I made my bed—a sheet of plastic and the sleeping bag—and prepared supper in the starlight. Cottonwood makes poor fuel, burns too fast and smoky; I scrambled up the talus under the cliff, where I had earlier noticed scrub oak growing, and brought back an armload of *real* wood. Over the red coals I hydrated my dehydrated meat and vegetables, stirring up a quick goulash. Well seasoned with salt and pepper and blowsand, nothing could have tasted better.

In the morning I went on down the gulch, leaving my pack behind, carrying only enough raisins and jerky for lunch. I was now below the seepline in the overlying Navajo Sandstone formation and the stream grew larger, fed by the many seeps and springs. I climbed around a gap in the rock, which had once been a natural bridge, passed several small waterfalls, came to a natural bridge still standing. This is a young one, geologically speaking, with plenty of room for enlargement. At present it is about big enough to drive, say, a schoolbus through. Clinging to the underside of the bridge are the mud nests of cliff swallows. Against one buttress, out of the rain, I noticed the smudge of an old campfire—as if some weary hobo, accustomed all his life to sleeping under bridges, had once spent a night here. I probably know the man.

There is a sensuous pleasure in walking through such a natural tunnel in solid rock, wading the stream that made it, standing in shadow and looking back through the bridge at a portion of the upstream canyon bathed in morning light—the sparkling water, the golden slickrock walls with their tapestries of desert varnish, the bright, acid-green leaves of the cottonwoods, the rosy pink plumes of tamarisk. From the cliffs far above I could hear the sweet clear falling notes of a canyon wren—a characteristic song in this land of stone and stillness.

What would it be like to live in this place? Could a man ever grow tired of such a scene? Someday, I thought, I will make the experiment. become an ancient bald-headed troglodyte with a dirty white beard tucked in my breechclout, starving on locusts and lizards, feasting from time to time upon fresh, lost Boy Scouts. Madness: of course a man would go mad from the beauty and the loneliness, both equally incomprehensible. But perhaps it would be—who can say?—a kind of blessed insanity, like the bliss of a snake in the sun, a buzzard in the air.

The canyon grows bigger, wider, wilder as it descends rapidly, in a series of jump-offs and cataracts, toward the Escalante. At one place a vast section of the north wall has collapsed quite recently, perhaps within the past century, and has tumbled in fragments the size of boxcars into the canyon bottom. The rockfall is clean, sharp-edged, free of all plant life, even lichens, and the floods have not yet had time to round off and polish the broken slabs that obstruct the creek. Now the footing is tricky, impossible for cattle and horses; I found steps chipped in the sandstone above the fall where some Mormon cowboy years ago had made a bypass for his horse.

More cascades, some of them fifteen to twenty feet high; at one point the only feasible route is down a log that leans against a shelf of stone on the canyon wall. Here the walls must be close to a thousand feet high on either side, and the sky is only a winding strip of blue. I walked around one more bend, hearing beyond the low steady roar of floodwater, and came to the master stream, the Escalante, swollen by the spring runoff—snow melt from the high plateaus eighty miles away.

The water was reddish brown, about ten yards wide and—as I discovered when I waded into it—knee deep in the main channel, with a strong current. There was no choice, however, but to stay in the water if I wanted to go farther; the river filled most of its canyon from wall to wall. I headed up the river; at least that way I was fairly sure I'd be able to get back to the mouth of Coyote Gulch in case the river should rise still more.

The going was hard against the current. I could feel the sand giving way beneath my tennis shoes at every step. I went only far enough up the river to see the great high window in the rock called Stevens Arch and to reach the mouth of Stevens Canyon beyond it. I had hoped to explore Stevens Canyon for part of its length, but the day was already more than half over. Reluctantly I turned back, waded the river, and walked the long winding rock-and-mud, sand-and-water way up the meanders of Coyote Gulch. Again I cooked supper in the dark—tired, hungry, happy.

On the third day, before returning up Hurricane Wash to my car, I decided to have a look at the weird world of rock and juniper that lies above and between the canyons. Upstream from the confluence of the wash and Coyote Gulch I found an egress from the canyon up a rounded knoll of sandstone. I kept climbing until I topped out on a bare ridge about halfway between Coyote Gulch and Escalante Canyon. From there I could see not only the dark gashes of the canyons below but also that hundred-mile-long monocline known as Waterpocket Fold, the snow-covered dome of Navajo Mountain, the Kaiparowits Plateau, and other mesas and plateaus east of Lake Powell and the Colorado River, some of which may have been—from where I stood—close to a hundred miles away by line of sight.

My immediate surroundings were no less interesting: a rolling terrain of largely naked sandstone shaped by ages of erosion and weathering into a surreal landscape of knobs, turrets, pinnacles, fins, potholes, and hollows. Parallel joints and seams in the rock lead to the canyon rims; cutting across these joints are a number of crevices too deep and perpendicular to descend, too wide to leap. A silent world of stone, and all of it—despite the fissured surface and the sculptured relief—united in one massive monolith of rock, ancient as the dinosaurs.

The area is not entirely lifeless. Some of the potholes contained water when I was up there, and in the water were threadworms, various kinds of larvae, and the curious freshwater crustacean, *Apus aequalis*—a holdover from the Paleozoic Era—much like a trilobite or miniaturized horseshoe crab in appearance. In the basins and depressions among the sandstone hummocks where sufficient sand has accumulated to form the basis of a soil, I found junipers and other plants, many in bloom: yucca, cliff rose (fragrant as orange blossoms), prickly-pear cactus, Indian paintbrush, princess plume, purple penstemon, globe mallow, sand verbena, and evening primrose.

I'd neglected to mark my trail and so spent a couple of hours searching for the way back down into Coyote Gulch. All of the sandstone hills look dismayingly alike when you're lost and my first two approaches to the canyon ended at rimrock. Each time I had to backtrack, circle, and try again; at last I found the one and only route down. Strange and beautiful though it is up in the slickrock wilderness, I had no wish to spend the night there without food and sleeping bag. Down in the canyon I returned to my pack, had something to eat, filled my belly with water, and began the twilight walk up Hurricane Wash toward my car and the road that leads back, not necessarily to home, but to where we all are now.

A Footnote to the Walk

It will come as no surprise to readers to learn that a place as serene, lovely, wild, and magnificent as the Escalante wilderness is not to be left undisturbed. Anything good can always be improved upon by a vigorous application of dynamite, bulldozers, draglines, dump trucks, and earthmovers, finished off with a thick, smooth coating of asphalt. Or so believes the corporate brain of any normal state highway department and its supporting combine of auto dealers, road contractors, construction unions, filling station operators, motel managers, and small-town Chambers of Commerce—all the homogeneous Babbittry of America who have not lessened one degree in avarice, myopia, or enthusiasm since Sinclair Lewis wrote his futile book more than forty years ago; who have, if anything, become even more vulgar, even more aggressive and destructive, in the present. Profit is their aim and pollution is their grossest national product. Pollution in every sense—and of every sense.

The Utah State Department of Highways conforms to the pattern devoutly. Where I walked out of the gateway of Coyote Gulch and into the deep golden canyon of the Escalante, under the blue window in the cliff called Stevens Arch, the state engineers propose to blast roadways through the canyon walls and join them with a huge bridge. The bridge will serve to complete a projected highway that will begin in Grand Junction, Colorado, cut across the southeastern corner of Utah, and terminate near Page, Arizona. (Proponents of the road estimate its total cost at $24 million; those who oppose it, at $90 million.) The 37-mile stretch of highway that would cross the Escalante area begins at Bullfrog Basin, northeast of the Canyon, and ends southwest of the Canyon at a place known as Hole-in-the-Rock. If constructed, this portion of the highway would lead, in effect, from nowhere to nowhere.

Hole-in-the-Rock is the possible site of a possible marina. Not a good site, because it is too small in area and some thousand feet below the end of the access road. (The gap varies with the fluctuation of the water level in the reservoir.) Although of modest historical interest, there is no development of any kind at Hole-in-the-Rock now. Nobody lives there and nobody has ever lived there; it is sixty miles by primitive dirt road from the nearest town (Escalante).

The other end of the proposed Escalante stretch of the highway (to be paid for, of course, out of federal, not state or local, funds) would lie in the place called Bullfrog Basin, a small and remote marina that is destined to be silted in and useless, some say in fifteen years, some say in thirty. Bullfrog Basin is accessible by paved road via Hanksville, from the north.

The estimated cost of the 37-mile highway from Bullfrog Basin to Hole-in-the-Rock is $6 to $7 million (although the road's opponents claim a cost of about twice this sum). The high cost of such a short auto artery follows from the sad fact that the planned route, which is the only possible route, goes against the grain of the landscape and at right angles, or crosswise, to the natural lines of travel in the area. To get from Bullfrog Basin (population about twenty souls?) to Hole-in-the-Rock (uninhabited, nothing there) the builders will not only have to blast through the walls of the Escalante and bridge the river, they will also have to lay part of the roadway on talus slopes of sand and tunnel or notch or cantilever a major section of the road through, or somehow over, the complex canyons and escarpments of the Waterpocket Fold. The road can be built; our engineers can do anything if we permit them. But this particular road would be extravagantly expensive, unavoidably destructive to the landscape, and conspicuously useless.

Perhaps the last objection is not important. Art is useless; this article is useless; moon trips are probably useless. Technicians have as much right to do their peculiar thing as anyone else, since it pleases them, however silly it may appear to others. Uselessness, then, is of little importance. The essential objections are the expense and the destruction.

The small businessmen of the Escalante area (and since the area is largely wilderness they are not many) think that any kind of public construction will benefit them in some way, so long as the cost is borne by the rest of us. But there are a few things that even scantily populated southern Utah needs more than nowhere roads—such things, for example, as new schools, medical services, better housing, agricultural aids, and pollution control. Since the money for the proposed highway is to come from the Economic Development Agency, it would seem appropriate to spend it on actual human needs rather than on one more facility for motorized tourism.

Misappropriation of public funds is bad enough, but the destruction of the landscape and the further bisection of the last remnant of the Glen Canyon wilderness—of which the Escalante River and its canyons is a part—is worst of all. Most of the people of southern Utah do not realize it but their most precious natural resource is not carnotite or potash or sand-tar oil, it is the grand, spacious, free, and sun-bright land itself—the whole primitive wonderland of canyon and mesa and desert. The living heart was Glen Canyon and the river, and someday they will both be restored to us. (I am not one of those who believe that the Powell reservoir is here to stay; I predict that Glen Canyon Dam, already obsolete as an efficient power-producing system, will be dismantled and removed by the more enlightened generations that are now coming up from the universities.) In the meantime we must treasure what little free space and open air still remains, save it from the road builders and the other varieties of commercial exploitation. The Escalante wilderness is the best we now have left of what was once something better, and it is federal land, public land, the property of all Americans. Its magic and beauty can never be reached by automobile because wilderness is the essence of it; but the canyons are open and available to all who are willing to make a little physical effort and who retain in their hearts some residue of that love for adventure that is, supposedly, a basic trait of our national character.

A Footnote to the Footnote

Detailed information concerning the trans-Escalante highway controversy can be obtained from:
(1) Information Office, State Highway Department, State Office Building, Salt Lake City, Utah 84114 and (2) The Escalante Wilderness Committee, P.O. Box 8032, Salt Lake City, Utah 84108

17

"A Song of San Francisco Bay"

by Malvina Reynolds

Malvina Reynolds received her Ph.D. in English from the University of California at Berkeley in 1939. Since that time she has composed and recorded numerous songs, the lyrics of which are generally social critiques combining humor, indignation, and irony. She has written three books of songs, Little Boxes and Other Handmade Songs, The Muse of Parker Street, *and a children's songbook,* Tweedles and Foodles for Young Noodles, *selections of which have been recorded in an album by Columbia Records. In addition to her affiliations with the American Federation of Musicians and the American Society of Composers, Authors and Publishers, she belongs to the Sierra Club and the Wilderness Society.*

The highway is laid as smooth
 as glass
For miles and miles and the
 cars can pass.
But the ant and the bee and
 the bush and the tree
Whose home it was are now
 exiles.
And the cars rush by for miles
 and miles,
To find a place where they
 can see
A plant, a bush, and a blade
 of grass,
And a ladybug, and a bee.*

*From "The Highway," words and music by Malvina Reynolds. Copyright © 1965 by Schroder Music Co., A.S.C.A.P.

Of the 270-odd miles of shoreline of San Francisco's great bay, some—not much, but some—is still in its natural state: grassy or lightly wooded hills leading down to a rocky border; marsh and tidelands green with reeds and salt grass, still alive with sea and shore birds and the infinite smaller creatures who populate such places. And if you wanted to see something like this, you might take a thirty-mile drive along the bay's northern shore—say on Highway 37—between Vallejo and San Rafael, with no towns between. If you'd been there before, a few years ago, you would remember a long stretch of secondary road a few feet above tideland level, with open marsh or farmland on both sides and the bay itself stretching to the south.

But now on that route you would find no bay, no marshes, not even farms for miles beyond Vallejo. Just the usual decorations of the modern highway—auto junkyards, hamburger stands, motels, small factories, billboards, gas stations—and the bay not even in sight. So you might take one of the unnamed roads that heads south to where the bay should be, and you could find yourself in the enormous grounds of the Kaiser Steel plant. Driving into the grounds you'd go over a bumpy road through an expanse of raw ocher earth, with a lonely bunch of reeds in a small pool at a culvert showing what had been here before the steel mill took over. As far as the eye can see, lines of flatcars on the spur track carry monster pipe probably destined to provide drainage for the new highways that are being laid everywhere you look.

This isn't what you were looking for, so you bump back on the branch road to 37, into the roaring raceway of cars, heading westward again. It is ten miles or more before you finally reach open fields and then the marshes, with reeds and salt grass and, for the first time this trip, the cool wind that blows from the bay and across the flats.

After a while you want to stop, but it isn't easy to turn off, with the heavy traffic averaging 50 miles an hour, and some people might wonder why anyone would want to stop here anyway. No garage, no hot dogs, no bar—the cars all tearing by, headed some place. But it happens there are white herons feeding not far from the road, and you hope to get them with your camera. They have become accustomed to the heavy trucks roaring by and the endless line of cars. But you have to come up easy with the camera—they are not accustomed to people on foot.

Seventy miles of wind and spray,
Seventy miles of water.
Seventy miles of open bay—
It's a garbage dump.†

†From "Seventy Miles," words by Malvina Reynolds, music by Peter Seeger. Copyright © 1965 by Abigail Music Co., (BMI).

I wrote the song "Seventy Miles" in March of 1965. For me, it marks the first time I became sharply aware of what was happening to San Francisco Bay—my bay, everybody's bay.

In a way, San Francisco, port to the coast trade and the Orient, takeoff point of the gold rush of the 1850's, commercial center of the mineral and agricultural wealth of the state, is everybody's city. Will Irwin's *The City That Was,* a little "requiem of old San Francisco," written in 1906, three days after the great earthquake and fire, opens with a quote from one Willie Britt, "I'd rather be a busted lamp post on Battery Street, San Francisco, than the Waldorf-Astoria." And Irwin himself describes the city of those days as "The gayest, lightest hearted, most pleasure loving city of the western continent." Fifty years later, Joseph Henry Jackson, book reviewer of the *San Francisco Chronicle,* opens his booklet tribute *My San Francisco* saying, "San Francisco is a great and greatly loved city."

Songs are written about San Francisco in every generation; visitors who come here go back home to pack their belongings and get out here to live. Whether they came with the covered wagons, with the railroads, with the gold-rush pioneers; as servicemen in many ways, as workers in war industries, or just as summer tourists running away from the muggy heat of other parts of the country—they settled here at last. Areas around the metropolises of the West Coast are among the fastest growing communities in the world.

And even if they don't get here to live, if they have only visited or if San Francisco is only a name to them, they feel a sense of identity with this city; it is part of the romantic tradition of America and the West. But I was born here. Grandparents on my mother's side were pioneers who left Lithuania in the 70's and, rejecting the East Coast, came directly to San Francisco. My father came alone from Budapest when he was fourteen. I was born south of Market Street in San Francisco—now an area of small industries and wholesale houses, but then a respectable lower-class section of wooden railroad flats, a few of them still standing among the warehouses and plants. I have lived in this city or near it most of my life. As the saying goes in this town, "Why should I go anywhere? I'm here already."

The bay and the hills around it are part of my natural environment, as they are of most of the three million who now inhabit the bay counties. Until about 1965, I had taken this great inland sea for granted. But all the time, it was being edged in on.

One of the maps in the office of the San Francisco Bay Conservation and Development Commission is a colored layout of the bay area showing the landfills that have been made since 1860. By 1900, when I came on the scene, the three coves that had sheltered sailing ships on the eastern side of San Francisco peninsula had been largely filled and built over. Later, while earning my degrees—and learning how to have fun—at the university in Berkeley, agents for farm land were filling about twenty square miles of mud flats and marsh in the northeastern arm of the bay; and under my eyes, as I crossed back and forth on the new transbay bridge, a whole new island had been laid in the shallows next to Yerba Buena Island, to accommodate first a world's fair, and then a Navy base. Some square miles on the eastern shore, south of Oakland, had also been filled.

In 1940, my husband Bud, a construction worker, was up to his ankles in mud, working on the shipyards that would be part of a continuing extension of the town of Richmond, jutting out from the bay's east shore. In 1942 and 1943 he worked on the Oakland Naval Base, which, with the adjacent Alameda Naval Air Station, thrust fill and docks for a mile or more into the bay alongside the East Bay bridge approach. At the same time the San Francisco airport, built on fill at the south end of the bay, was granted an extension of submerged lands. Ten years later, 7,000 acres of tideland would be granted to Oakland for its airfield. Two beautiful small bays on the northwest shore, in Marin County, had been converted from jewels reflecting Mount Tamalpais and the wooded foothills into flat, dirt-filled areas with a few buildings for decoration.

Free fill supplied by the Army Engineers from the dredging of the bay bottom was part of the material used to make these continual fills. More came from hills that were being leveled for factories and housing sites. The most offensive was garbage. At this writing, the city of San Francisco still dumps its tremendous daily load of garbage into the southern end of the bay, in spite of a valiant fight by people of the city of Brisbane, on the face of San Bruno Mountain and overlooking the flats, against the offense in their front yard. There are more than thirty refuse disposal sites in and at the edge of San Francisco Bay, from Martinez and Pittsburg on the northeastern arm to Palo Alto and Alviso at the south. And the discharge points for sewage and other fluid wastes, domestic and industrial, are about three times as many.

Because the mountain formations of the West Coast run generally north and south, there are very few good harbors on the Pacific. San Francisco Bay is the largest and finest south of Puget Sound, a water area at mean high tide of more than 435 square miles, almost completely protected from the sea except for the strait on the west called the Golden Gate. The bay itself lies in what was a long valley, extending north and south. Into its northern arm empty two big rivers of the Central Valley —the Sacramento, flowing south, and the San Joaquin, flowing north.

More than 400 square miles of open water. But when California was admitted to the union, before any "sanitary landfill" had been dumped into the bay, its area was more like 700 square miles. If you tell the average San Franciscan that the bay is one-third smaller than it was when discovered, he will be startled. And if you tell him that a good deal of the rest of the bay—and I am talking of water-filled bay as far out as the deepwater boating channels—is city, county, state, or privately owned real estate, marked into lots and changing hands at ever rising prices, he will question your sanity. But it's true. Mel Scott, Research City Planner commissioned in 1963 by the Institute of Governmental Studies of the University of California at Berkeley to do a study on the future of San Francisco Bay, tells of the ownership of these submerged tide and water lots in the bay: "Almost the entire waterfront of Sausalito has been consolidated by a single syndicate. Just seven owners now control the submerged properties in Corte Madera Bay, among them the Marin Title Guarantee Company, the Utah Construction and Mining Company, the Wells Fargo Bank and the City Title Company. . . . On the eastern side of the bay the Santa Fe railroad has title to almost all the privately held tide and submerged lands from Richmond to Oakland, and in Richmond alone it owns 1,156.13 acres. The Standard Oil Company possesses more than a thousand acres of offshore properties in Richmond, and claims title to another 640 acres." And so on and on. I sat on the deck of my friends' pleasure boat in the Sausalito Yacht Harbor on a pleasant Sunday afternoon, and they pointed to a section of the bay, far from shore, where yachts and trimarans were sailing by. "A lot out there," they said, "just changed hands at a price of $600,000."

"For eleven decades," says Scott, "the state government has tended to regard the bay as property rather than as a great natural resource to be safeguarded." While the state still owns the greater part of the northern arms of the bay—Suisun and San Pablo bays—it is in possession of less than half of the main body of the bay.

This process of local and private occupation of underwater real estate and tidelands began even before California became a state, with an edict issued in 1847 by Brigadier General Stephen W. Kearney, military governor of the territory, for the sale of submerged lands in Yerba Buena Cove at public auction—"proceeds for the benefit of the town." The cove exists no longer, but is filled from Montegomery Street, San Francisco's Wall Street, to the Embarcadero, the wide avenue along the bayside docks. It now supports a sizable part of the downtown city.

They say the filling process began, perforce, when ships of the '49 Argonauts were abandoned in the cove, a whole fleet that had come around the Horn and anchored there—the first no-deposit, no-return throwaways in the state's history. Some of them sank and are part of the underlay of tremendous office buildings; others were used as temporary dwellings and business places in the exploding town of the mid-century, and the area around them was filled for foot and carriage passage. The ship *Apollo* became a saloon; the brig *Euphemia* served as a prison ship.

At any rate, the process of private ownership of the bay was made official by the Army officer who is generally forgotten even by those who use busy downtown Kearney Street, which is named for him. And the process has continued under confused authority, until we have the situation described by Mr. Scott, with much of the bay now in title deeds. The agencies trying to hold the line against filling of the bay have to battle these holdings, foot by foot.

At first thought, to a tenderfoot like me, much of the bay would seem to be of questionable value in its native state. A good deal of it is quite shallow even at high tide, and the U.S. Army Corps of Engineers, with the engineer's practical eye, describes 248 square miles of tidelands as "susceptible of reclamation."

You will excuse a poet's digression here. I am sensitive to words like *reclamation* and *development* because they are not, as they would seem to be, impersonal and descriptive. They are highly charged with propaganda. My dictionary says of "reclaim": "to bring back to a useful, good condition," and two common uses of the word are given in the quotations that follow: "The farmer reclaimed the swamp by draining it." "Henrietta reclaimed him from a life of vice."

We are to believe that the dumping of waste and garbage into the edges of San Francisco Bay, the covering of it with the bedrock of ravaged hill, is a reclamation.

The building of housing and businesses would at first thought fulfill the idea of "development," which Webster's Twentieth Century Dictionary describes as "the causing to become gradually fuller, larger, better."

But the development of the bay region has one immediate, adverse effect. In a crowded part of the world, it puts people in residences and in factories where there were no people before. All the harmful results of overpopulation are immediate—increased strains on waste disposal facilities, increased appearance of pollutants in air and water, increased demands for highways and parking facilities, increased psychological damage from crowding.

But the shallows and mud flats in their natural use have values of their own. It does not take an expert to know that open space is a great value in itself. When it carries bay and ocean breezes, sweetened with oxygen and natural moisture, it is even more precious. And the mud flats and marshes are in themselves producers of oxygen. Mud algae, exposed to sun and water, produce oxygen for air and water. So do the marsh plants, the tules and the marsh grasses. The latter are very rich in food value for the sea and bay creatures that feed on them as the plants die away. Cord grass, one of the marsh plants, has seven times the food value of an equivalent acreage of wheat, and in the cycle of life that goes on in the waters of the bay and the ocean, these food values insure a rich crop of fish and other sea food for the multimillion-dollar sport and industrial fisheries.

I'm sorry to find myself talking in money terms. But every agency that fights for preservation of our bays, rivers, and forests finds itself having to appeal in such terms to the governing bodies—city, state, and national—that have these decisions in their hands. Otherwise, they can be accused of "catering to the bird watchers."

The bird watchers—there's
  a laugh.
Their idea of having fun,
Sitting on a hillside under
  the sky,
Sensing the trees and feeling
  the sun,
Watching the birds who nest
  and fly,
Watching the castle clouds
  go by,
Watching the flowers, watching
  the bee,
When they could be sitting at
  home with a beer,
Watching T.V.

Besides being the home of many local characters—sea gulls, pelicans, herons, cormorants, and loons—San Francisco Bay is the stopping station for hundreds of thousands of birds on the Pacific flyway, a run that extends from South America to the Arctic Circle. The marshes and waters of the bay, where the water is less than eighteen feet deep at low tide, are the principal feeding, resting, and wintering grounds for seventy-five different species of water birds that visit and live here. Mussels, clams, snails, worms, and insects are the main of their diet;

the deep waters of the bay are their playground and their shelter during storms at sea. The bay has a great fish population, too, and many species that live in the sea come to these shallows to spawn. The young feed on the marsh and mud flat life, and vice versa.

There are about fifty square miles of marshland remaining around the bay, sixty-five square miles of tidal flats, and seventy-eight square miles of land at the south end of the bay that are used for salt production—commercial, indeed, but still an open area and one that sustains certain water creatures adapted to the concentrations of salt in the drying pools.

But if, in your scenic drive, you missed the way to Highway 37 and came to the new bridge over the Napa River, which feeds into the bay from the north, you would find inroads into the open areas—new highway being laid over the marshland; raw yellow earth over a wide swath of what had been reeds and pools and small sea creatures. A poet sitting by the driver of the car might have scribbled in his notebook:

> Goodbye to the reeds and the
>   little fishes,
> And the shouting boys who
>   play under the bridges.
> Raw earth dug from the doomed
>   hillside
> Pours over the marshgrass,
>   child of the tide;
> The living wind that blew
>   from the bay
> Now carries the fumes of the
>   swarming highway.

And at scores of places on the bay shore, and notably at the south end, eight thousand tons of assorted garbage—trash and the stones and waste of demolished buildings—are daily dumped into the bay.

We are generally an affluent and wasteful people, and each one of us produces his four pounds of discards a day—not to mention sewage and the products of industrial waste that our consumption of stuff requires.

What to do with all this? I don't know. I only know that a civilization that can send men to the moon can certainly find other ways of handling this problem than by destroying a city's greatest natural resource for a dump. I read somewhere about a European city that consumes its waste under very high heat, and presses the resulting slag into building blocks. What a great idea! If this is so, then maybe it doesn't have to fell its forests for building materials, either. Apparently there are many kinds of engineers in the world.

The tone of my discourse, so far, would indicate that I favor pelicans, wild geese, striped bass, and sea snails over human beings. I would hate to be pushed too far on this question, but it is really not relevant, because people and pelicans alike live much better lives as one in the life-chain with marsh grass, ladybugs, forests, robins, wild blackberries, and shrimps. If more people were packed into the space already marked off in the bay as real estate, they'd all be worse off than if they had never come here in the first place. California is a tremendously large state, and there are millions of acres that could absorb any normal increase in population. That is, if we weren't all city-oriented, which is another question.

The great pulse of the ocean tides sweeps into San Francisco Bay twice a day, and ebbs out, to create a tremendous circulatory system, which cleanses the bay waters, aerates the waters and mud flats, and feeds the creatures who depend on this tidal change for their environment. Meanwhile, fresh water from the Sacramento and San Joaquin rivers flows into the north end of the bay, sometimes over the salt water beneath. In the central area of the bay the ocean tides can often be seen moving in over the underlying fresh water, and underneath are tremendous currents caused by these pulls and drifts, as well as by the contours of the bay's edges and bottom. The oxygen content of the water is affected by these movements, and on this content depends the ability of the waters to handle and convert the outflow from the cities, to prevent the waters from becoming lethal to wildlife and people. Every reduction in the area of the bay works against this function, and about now, the margin of safety is pretty small. Four and a half parts of oxygen per million of water are necessary for the survival of the fish and other creatures of the bay waters. It now stands at an average of something like seven to nine and a half parts per million—and the garbage trucks and dump trucks are running down the Bayshore Freeway all day, every day, while the engineers offer free fill to all who want it.

According to the San Francisco Bay Conservation and Development Commission, if all the relatively shallow parts of the bay shown by the U.S. Army Corps of Engineers to be "susceptible of reclamation" were filled (and, as we have seen, about 25 per cent of the bay, including most of the shallow areas nearest the shore, is claimed by private owners), the bay would consist of only 187 square miles. In some places it would be little more than a river. The rest would be what Harold Gilliam, feature writer of the *San Francisco Chronicle,* calls, in an article in the Sunday magazine of March 7, 1965, a "fuming flat." The local provincial patriots among the conservationists fearfully describe the resulting place as "another Los Angeles." I am something of a neutral in this conflict between cities, but I think they may have something.

Anyway, what I think of the fill projects—and every county, every owner of submerged land in the area, has such projects in blueprints—opens the song, "Seventy Miles."

> What's that stinky creek
> out there,
> Down behind the slum's
> back stair,
> Sludgy puddle, sad and
> gray?
> Why man, that's San Francisco
> Bay!

Harold Gilliam, quoted above, is one of the great naturalists of the Pacific Coast, and a favorite columnist on San Francisco's only major morning paper. His work on San Francisco Bay, published in 1957 by Doubleday, is a learned and poetic tribute to what he calls "the incomparable harbor." Each section of the book opens with a lyric painting of a particular aspect of the bay, its hills, and its sky.

"Clouds hang low over the vast amphitheater of the bay and its shores like a pavilion roof supported by mountain pillars. The scene is a monochrome of grays of infinite shadings from silver to near-black. The morning air is still, and the bay is flat and glassy—a clean scroll ready to be written on by the winds, the fogs, the rain, and passing vessels.

"A freighter leaves a spreading V-shaped wake of waves and ripples sharply etched on a surface of gunmetal gray. A tug crosses the ship's wake and sets up a conflicting series of wave lines. . . ."

And, in another area of literature, John Phillips, songwriter member of the Mamas and the Papas, sings:

> If you are going to San Francisco,
> Be sure and wear some flowers
> in your hair.*

Too bad, too bad, that we have to talk of garbage, and the proposed tearing down of a mountain (the only one in the southern area of the bay, San Bruno Mountain) for fill to eradicate another swath of this great inland sea. It is too bad that we have to talk of sewage outlets, of "industrial parks" and freeways and housing developments projected for the precious wild spaces that still remain and and even for the deepwater areas within the bay.

But I am not the only one by a long way who "came to" in the early sixties as to what was happening to my bay, I discovered, when I looked around, that many conservation organizations were actively concerned, that leading citizens of the area had formed a Save the Bay Committee, that the counties adjoining the bay had realized that some measure of joint planning was necessary to prevent the disaster that they were all beginning to forsee otherwise. There is now an Association of Bay Area Governments, a Bay-Delta Water Quality Control Program, a Citizens for Regional and Recreational Parks, and under the aegis of the state legislature, a Bay Conservation and Development Commission. There is a moratorium on filling—not complete, as we have seen, but a hopeful start—while BCDC prepares and presents its recommendations. The concept of development is included, granted—but at least some of this development seems to be in the direction of rescuing the few miles of publicly owned bay shore for public parks, adding to these miles by purchase where it's possible.

The eastern arm of Richardson's Bay, one of the bays within the larger bay, was recently threatened with development as a private residential marina, but conservationists raised a quarter of a million dollars, bought the place, and then turned it over to the Audubon Society as a wildfowl sanctuary.

Though there are constant pressures for easy airfield extensions, building sites, highways, and dumps into this great open bay's shallower areas, there is hope that most of it can still be rescued. I had occasion to engage in a similar conflict in 1964, when the state Highway Commission (autonomous and arbitrary until it has become a statewide scandal) was determined to run a freeway through Golden Gate Park Panhandle—a beautiful approach to the city's largest and most famous park —a park that graced and gave play and open green area to a heavily populated section of the city. It was only a small group of citizens that initially opposed this move, and many who were aware of the strength and resources of the interests favoring the freeway plan thought they didn't have the hope of a chance. But more and more came in to help the little band of conservationists. The *Chronicle* backed the campaign, and called a big rally in the park's Polo Grounds. Even I was called on to come and sing my song "Little Boxes," which I was glad to do. But just before the rally I also wrote "The Cement Octopus," and I sang that too. At this writing, after several years, the Panhandle still remains a park. A year after the first victory, a little group of us met in the Panhandle in the rain to celebrate and plant a small new tree there among the giant conifers and eucalyptus that had been planted by John MacLaren, who created the park in the preceding century.

And talents and dedication of all kinds rally in defense of the bay. The Bay Area Photographers do a whole season's show on the bay, and contribute prints to the Save the Bay Committee. Junk sculptors construct art works on the east bayshore flats, littered with flotsam, tires, cans, and other trash, as a sort of mute plea for something creative, some joy for the eye, to come out of this senseless ravaging of the bay. The songwriters sing of San Francisco, and the patriarch of folksingers, Pete Seeger, in his Columbia album devoted to conservation (the great conservationist Justice Douglas does the album notes for "God Bless the Grass") sings our song, "Seventy Miles."

What's that stinky creek
  out there,
Down behind the slum's
  back stair,
Sludgy puddle, sad and gray?
Why, man, that's San Francisco
  Bay!

Chorus:
Seventy miles of wind and spray,
Seventy miles of water.
Seventy miles of open bay—
It's a garbage dump.

Big Solano and the Montecell',
Ferry boats, I knew them well,
Creak and groan in their
  muddy graves,
Remembering San Francisco
  Bay. *(Chorus)*

Joe Ortega and the Spanish
  crew,
Sailed across the ocean blue,

Came into this mighty Bay,
Stood on the decks and cried,
  Ole!" *(Chorus)*

Fill it there, fill it here,
Docks and tidelands disappear,
Shaky houses on the quakey
  ground,
The builder, he's Las Vegas
  bound. *(Chorus)*

"Dump the garbage in the Bay?"
City fathers say, "Okay.
When cries of anguish fill
  the air,
We'll be off on the Riviere."

Chorus:
Seventy miles of wind and spray,
Seventy miles of water.
Seventy miles of open bay—
It's a garbage dump.

We are working out here to guarantee that this song becomes obsolete—a curio.

18

"A Soviet Plan for Nature"

by Innokenti P. Gerasimov

A 1926 graduate of the University of Leningrad, Soviet geographer Innokenti P. Gerasimov works in the field of the genesis, geography, classification, and mapping of soils. His particular interest involves the interaction of endogenic and exogenic forces and the relief-forming role of the movements of the earth's crust. Elected Academician in 1953, Gerasimov is presently Director of the Institute of Geography of the USSR Academy of Sciences. During the last few years, he has been actively concerned with the practical application of modern research methods in geography. Professor Gerasimov is the author of several hundred articles on geomorphology, soil science, paleography, nature protection and transformation, and the restoration of natural resources.

The transformation of nature in the Soviet Union has a purposeful and scientific character. The Soviet Union was the first country in the world to initiate government measures for transforming nature over extensive areas for the benefit of the national economy. These measures are included in general outline in the Program of the Communist Party of the Soviet Union. They constitute an important part of our economic plans and have repeatedly been given concrete shape in decisions of plenary meetings of the party's central committee. Universal interest has been aroused by the way in which the state has tackled major scientific problems connected with the natural environment. The tremendous importance of rationally exploiting and preserving natural resources is recognized everywhere.

Because of industrialization and the rapid growth and concentration of population in large urban centers, the deterioration of our natural environment (the biosphere) is proceeding throughout the world at an ever increasing pace. This results in developments that endanger the well-being of mankind—the pollution of the atmosphere and soil beyond all permissible limits with domestic and industrial refuse, a shortage of land and other natural resources owing to their irrational use, the irreparable destruction of natural ecosystems, and the complete extinction of many species of living organisms.

The ever increasing extent to which man is interfering with his natural environment makes it essential, in my view, to prepare a general scientific plan for the transformation of nature. In contrast to the capitalist countries, the USSR, with its planned socialist economy, should be able to exploit nature in a purposeful, rational manner.

The tremendous hydroengineering achievements of the Soviet Union, which have greatly increased its power potential and helped to bring about basic improvements in important waterways, are well known. The rapid development of the power industry has been made possible by redirecting the natural flow of rivers, mainly in the European parts of the country, in the basins of the Volga, Dnieper, and Don, and to some extent in the big rivers in Siberia and Central Asia.

However, these solutions to some important economic problems have led to serious consequences that were not anticipated. The full extent of these unforeseen consequences has not yet been revealed, but what is already known should undergo comprehensive scientific and economic analysis. This is necessary in the first place in order to forecast the impact of future hydroengineering schemes.

Some of the consequences of the big hydroengineering developments of past years were unavoidable. One result was the flooding, by artificially created reservoirs, of tracts of agricultural and forest land covering many millions of hectares (one hectare=2.47 acres). Even larger areas have been subjected to indirect effects of the reservoirs—swamping, salinization, disintegration of banks, and soil erosion. These processes, which spoil the land in areas bordering on the reservoirs, are continuing, thus requiring urgent work on drainage, the strengthening of banks, and land improvement.

Under the influence of such reservoirs, the total water exchange in the hydrographic basins affected by the alteration of the flow has been greatly reduced. This reduces the capacity of many rivers to rid themselves of the growing quantities of industrial and domestic waste, and a gradual pollution of the river occurs. As a consequence, the water supply problem, serious as it already was in many industrial regions, has been further worsened.

Urgent and complex engineering measures have had to be taken in order to control and process sewage and to make use of a closed water cycle. So far, these measures have failed to overcome all the water supply difficulties.

The regulation of the natural flow of rivers has increased the losses of moisture through evaporation from reservoirs. This is particularly great in the steppe and desert areas and has changed the subterranean water-feeding conditions through the infiltration of water from the rivers. All these changes have not yet been sufficiently investigated.

Far more obvious are the deleterious effects of hydroengineering developments on the fishing industry in practically all the internal basins. The resultant disturbance of the natural behavior of the rivers, their deltas and estuaries (especially in the case of the Volga, Dnieper, and Don basins), has led to very heavy losses in the fishing industry. Furthermore, various special engineering and economic measures—the building of fish-breeding farms, artificial fish hatcheries, and so on—have had to be taken to combat these effects.

Major new hydroelectric schemes are being planned in the Soviet Union. There can be no doubt that their implementation will bring about many changes, like those already described, in the natural environment of vast territories, but on an even larger scale. With accumulated experience to fall back on, it should be possible to anticipate some of the probable changes and to take measures to prevent or at least mitigate their harmful effects. This, however, is not enough for a complete and reliable forecast to be made of all the complex and interconnected natural phenomena that may arise and develop as a result of large-scale hydroengineering work.

A comprehensive account of all the economic gains and losses involved in a project must be part of the forecast. In addition to the direct building investments, the revenue to be derived from the electricity produced, and the increase in agricultural production (with the development of irrigation), the calculations must cover indirect investments and losses resulting from the disturbance of natural conditions and the measures required to forestall unfavorable changes in nature. In forecasts of this kind all natural resources must be given an economic, monetary assessment.

How important this is may be seen, for instance, from the physicogeographical analysis made of the natural phenomena of West Siberia and the forecast made of the changes likely to occur if the Lower Ob hydroelectric station is built. An analysis has shown how absolutely impermissible it would be to build even a low-pressure dam in the part of the Ob immediately above the delta. Any head of water created there, this forecast has revealed, would have a disastrous effect on the general water regimen of the West Siberian plain and would increase its boggy nature, which is already hampering the development of this territory.

On the other hand, regulating the flow of the Ob and the Irtysh, and their tributaries in the mountainous parts of their catchment areas, would have a beneficial effect on the water regimen of West Siberia.

Scientific forecasts of this kind must become an indispensable part of all design operations. In order to do this, it will clearly be necessary to go beyond the solution of purely hydroengineering problems.

Our total cultivated land amounts to 225 million hectares. Nearly three-quarters of this area is situated in arid areas and in steppe and forest-steppe zones, about one-tenth in desert, and only about a quarter in forested and adequately moist zones. As regards its natural conditions, therefore, agriculture in the USSR differs markedly from that of the United States and of many countries of western Europe, the difference being fundamentally to our disadvantage.

What I am referring to is the periodic, and in some cases systematic, shortage of moisture in the soils of the arid zones. This is largely why our average yields in areas without artificial irrigation are lower than in many countries with more favorable atmospheric moisture conditions.

An increase in the natural supply of moisture in cultivated soils, together with the development of artificial irrigation, will help to step up and stabilize crop yields in the Soviet Union. It also will produce the maximum results from the mechanization and chemicalization of agriculture, carried out jointly on a wide scale.

No less than two-thirds of the total cultivated land in the USSR, or nearly 150 million hectares, needs more moisture, while between 50 and 60 million hectares are subject to erosion. Yet despite the trmendous importance of the measures needed to create a favorable microclimate, increase the moisture content of the soil, and combat erosion in the arid zones, these measures are not yet being carried out on an adequate scale. Much could have been done in this connection at a minimum cost by the collective and state farms themselves.

In order to make our agriculture more productive, I believe it is necessary to draw up a general scientific plan of radical measures designed to improve the microclimate of territories, increase the natural moisture content of fields, and eliminate completely the destructive processes of erosion.

Important to the development of agriculture is the draining of boggy land and the irrigation of desert and arid land. Artificial irrigation sharply increases the biological productivity of land and makes agriculture more stable. But it demands the greatest care, and the fullest account needs to be taken of the specific conditions of the areas being irrigated. Underestimation of the importance of this leads to dire consequences.

The total irrigated area in the Soviet Union has increased greatly under the Soviet government, and now amounts to some 10 million hectares. This was achieved through the systematic engineering reconstruction of old irrigation systems and the irrigation of new areas in Central Asia, southern Kazakhstan, and Transcaucasia. This rapid expansion of irrigated farming has created new opportunities for agriculture in our southern republics.

However, the economic benefits of this extensive irrigation program have not been as great as they might have been because of secondary salinization of the irrigated soils. Secondary salinization has now spread to a substantial part of our irrigated land and has reduced crop yields. Moreover, it sometimes makes irrigated land unusable.

The local population, as a result of long experience, has developed methods of combating secondary salinization—periodically washing the salinated soils, extensively mixing them, and not cultivating them to excess. However, the engineering reconstruction of the irrigation systems carried out under the Soviet government requires that the land be cultivated much more intensively. As a result, there has been extensive secondary salinization of soils in many irrigated regions.

Age-old experience of irrigation farming, coupled with years of scientific research, now makes it possible to take into account all the basic factors in the development of secondary salinization and to work out a set of hydrotechnical and agricultural improvement measures capable of arresting and ultimately eradicating this process. However, these measures are only being applied to a fraction of the affected areas.

A special prognosis zoning of the entire irrigated (and boggy) territory of the country should recommend a water-and-salt regimen and determine methods and rates of watering and draining, which will insure the maximum fertility in the irrigated territories, and either arrest or completely eliminate secondary salinization. The drawing up of these forecasts would be an important part of a general plan for the transformation of nature.

Soviet scientists have proposed and substantiated the principle of the extended reproduction of all renewable natural resources. This is aimed at fully meeting the economy's growing demands for these resources and at bringing about a simultaneous and steady increase in the resources and their quality.

For instance, an over-all survey of land resources has led to the conclusion that the best land in our forest-steppe and steppe zones, totaling 230 million hectares, has been exhausted or is nearing that state from the standpoint of agriculture. Along with steps to preserve and enhance the natural fertility of these areas, any further extension of the arable area must be accompanied by large-scale improvements to raise the natural fertility of the land being brought into cultivation.

A comprehensive register and assessment of all the country's land resources is needed as a basis for all practical measures in agriculture. Unfortunately, no such register exists yet, and the work of compiling one is proceeding far too slowly.

Another important economic problem is the rational utilization of natural fodder resources, based on their improvement and reproduction. The Soviet Union's pasture and grazing areas and areas suitable for haymaking total 700 million hectares. Half of this area is made up of tundra grazing land of low food value, suitable for deer. However, many other natural fodder regions in the USSR are also marked by low productivity, due to insufficient efforts to better them by land improvement (draining, watering, and clearing of scrub).

Our scientific institutions have put forward many valuable proposals for improving the productivity of natural pastures and grazing land. The rational use of the natural fodder resources would open promising prospects from the economic point of view. Estimates show that if the necessary measures were carried out in the desert grazing lands of Central Asia, some 25 to 30 million sheep could be kept there —many times more than the present figure.

We have in the USSR a vast forest area amounting to 800 million hectares—a forest area more than five times greater than the corresponding area in the whole of Europe and nearly double the size of that of Canada and the United States. We rank first in the world in timber felling.

With their immense annual increment, the impression might be gained that our forest resources are not yet being fully utilized. A more detailed examination, however, leads us to a completely different conclusion. Our forest resources are distributed very unevenly over the country. This is due not only to the diverse natural conditions (the absence of forests in the southern regions and arid areas) but also to various degrees of deforestation. In the USSR there is a thickly forested zone (the northeast of the European part and of the Urals, western and eastern Siberia), along with a medium-forested zone (the northwest, the northern Urals), and a sparsely forested zone (the center of the European part of the country). However, the timber organizations pay little attention to the geography of forest resources. As a result, more felling is done in the more accessible, though less densely forested, areas, while less takes place in the heavily forested regions. Consequently, the forests are decreasing where there is excessive felling, and the composition of forests is deteriorating.

The existing timber system—felling and regeneration—and also timber-dressing methods are in need of serious improvement.

The principle of the all-round utilization of forest resources, based on their extended reproduction, has been universally accepted in theory. It is a long way, however, from being applied in practice, and our forest resources continue to be depleted of their most valuable parts.

The history and character of the exploitation of wildlife in the USSR have provided a wealth of material on the extended reproduction of natural resources. During the Soviet period a great deal has been done to restore the numbers of valuable animals by temporary prohibitions and regulation of commercial hunting and trapping, and by the enrichment of the fauna with the introduction of new species, often with remarkable results. A case in point is the restoration of the beaver and sable populations and the breeding of muskrats, mink, and other fur-bearing animals.

Animal breeding has now become an important industry. Our nature reserves have made it possible for many rare species of animals to survive, while others (such as saiga and elk) have increased in numbers.

Despite the losses that hydroengineering schemes have inflicted on the fishing industry, numerous measures have been worked out and implemented to breed fish in artificial conditions and acclimatize them in various river basins. Gray mullet, for instance, have been taken from the Black Sea and successfully acclimatized in the Caspian and Aral Seas, where they are increasing in numbers. The Pacific salmon has been acclimatized in both the Barents and White Seas.

The question may legitimately be asked: What is it in our society, with its consistent progress in all spheres of life, that interferes with a rapid advance in such an extremely important field as the rational exploitation of nature? The answer is far from simple. We must begin by admitting that the erroneous idea that natural resources are inexhaustible has taken deep root, and that a dangerous tradition has developed whereby the gifts of nature are used in a feckless and uncontrolled way. And the answer must conclude by noting that our scientific knowledge has not yet reached a level at which it can insure faultless control by man over the extremely complicated natural processes existing in our environment. It is clear that the foundation for such rational control must be scientifically accurate forecasts of the transformations in nature that will follow from human intervention.

We have already referred to the preparation of new hydroengineering schemes. Many of these provide for fuller regulation of the flow of the biggest rivers by constructing huge dams and reservoirs, and by large-scale diversions of waters from one river basin to another. Diversions of this kind have long been considered necessary in the European territory of the USSR—from the basins of northern rivers and lakes to the southern river systems (the Kama, Volga, Dnieper, and Don basins). Apart from supplying water for industrial regions, this diversion would increase the flow of water into the Azov and Caspian Seas, improve conditions in the lower reaches of the Volga and the Don, and facilitate the irrigation and watering of arid land and grazing areas in the southeast.

For a long time it has been proposed that part of the waters of Siberian rivers should be diverted to Kazakhstan and Central Asia, both to maintain the water balance of the Aral Sea and to promote irrigation schemes in the Central Asian plains.

The scientific and technical plans drawn up by our hydroengineering experts must, apart from the necessary technical estimates and designs, be based upon complex scientific forecasts of the long-term influence on the natural environment.

Take, for example, the problem of the regulation of river waters in the European territory of the USSR and the diversion of part of the waters from the northern basins to the south. Work on this problem must start with estimates of the probable changes in the water balance of those river systems and lake basins from which, and to which, the water is to be diverted.

Next, the various channels that will divert the waters and the technical requirements for making those channels will have to be studied. Forecasts of all changes in the water regimen of the river systems and lake basins involved, and also the flooding of adjacent areas, must be worked out at the same time. On the basis of these forecasts the various technical, land improvement, and economic problems will have to be solved, and measures taken to mitigate or eliminate the adverse effects of hydroengineering projects on the natural conditions.

All this, however, is only the first stage in a scientific approach to the problem. Since transformation of the flow of rivers on a very large scale is involved, the impact on the entire water exchange of the Russian Plain will be great. It will affect not only the water regimen and balance of many rivers and lakes but also the feeding conditions and the resources of ground and subterranean waters, including artesian wells.

It will also be important to forecast the amounts of heat and fresh water that will be lost by the White and Pechora Seas, the Gulf of Finland, and the Gulf of Riga, and the way in which these losses will affect the heat and ice regimen and the biological productivity of the sea basins.

The next stage will require an even higher standard of prognostication, connected with the analysis of the general geographical conditions involved in the formation of the water balance in the European part of the USSR. In the relatively recent past, practically all of the northern half of the Russian Plain was densely wooded, while its southern half was in the process of being slowly and unevenly brought into cultivation. Over the past 100 to 150 years the forests have been so intensively cut down in the center and northwest that little of them now remains, while the steppe area has been completely ploughed up. A new system of moisture circulation with a rapid and extensive runoff has been established.

Twenty years ago a start was made under state auspices on planting protective woods and forest shelter belts in the steppe and forest-steppe regions, together with the application of other methods for the retention of the surface flow. Later, large-scale drainage, afforestation, and other measures were taken.

All this could not fail to have a profound effect on the general circulation of moisture in the Russian Plain. The forecasts made under the new conditions of moisture circulation may enable us to adopt a completely new approach to the problem of determining the regimen and balance of the artificial reservoirs and to take a new look at the demands of irrigation and the possibilities of utilizing underground waters. Such a far-reaching prognosis could help us revise the entire system of water-economy estimates now in use, as well as the hydroengineering construction schemes.

Another recent, acute problem is the future of the Aral Sea. Ever increasing use of the waters of the Amu Darya and Syr Darya for irrigation purposes is already lowering the level of the Aral Sea and changing its characteristics. This process is bound to be further accentuated. But the problem we are examining is even broader. It consists in this case of the transformation of the entire moisture circulation between the mountainous part of Central Asia and its desert plains.

Special investigations must be made of the extent to which this general moisture circulation is self-contained and of the possibilities for its intensification—through artificial regulation of glaciers, interception of rainfall and snowfall in the mountains, and an increase of evaporation in newly irrigated areas and subsequent transfer and accumulation of moisture in the mountain ranges.

The role of the underground flow of mountain waters must also be cleared up. Perhaps it is precisely the extensive evaporation of this underground flow that serves as one of the main factors leading to the accumulation of salts in the soils of irrigated oases in Central Asia. Thus, the problem of the transformation of nature in Central Asia and south Kazakhstan by full use of the river waters for irrigation purposes should now be subjected to a new and comprehensive analysis.

In eastern Siberia work has started on the world's largest hydroelectric station. This involves the construction of dams across the Yenisei, the Angara, the Lena, the Vilyui, and other rivers. These schemes are encouraged not only by the technical and economic conditions but also by many of the specific natural conditions of these territories.

The need arises, however, for a scientific forecast of the general effect this hydroengineering project will have on the natural conditions of the regions adjacent to the rivers and the Arctic seas, for Siberia's mighty rivers exercise a great influence on the permafrost regimen of their valleys. And while the total share of this flow, as compared with other sources of replenishment of the ocean waters, is quite small, its role in the ice regimen of all the large ocean gulfs along the northern shores of the Arctic is great.

Consequently, the people engaged in designing huge hydroelectric stations on the big Siberian rivers should carefully weigh all the possible climatic effects, as well as the effects on the rivers themselves, the catchment areas, and the seas washing the northern shores. All this could affect not only the forestry and fishing industries in these areas but also the future of the Northern Sea Route, which has been developed at the cost of such heroic efforts.

In present-day conditions the preservation of nature is inseparable from its transformation and from the rational use of its resources.

To preserve nature in an undisturbed or slightly altered state, a territory must be declared a reserve in which no economic activities are permitted. The creation of a system of reserves in the Soviet Union was initiated by Lenin's decree on "The Protection of Nature," issued in 1921. Today there are more than 80 reserves in the USSR with a total area of 6,400,000 hectares. Since their establishment they have accomplished a great deal in the study and preservation of nature. They have helped not only to preserve typical tracts in their primary state but also to save many species of animals that were on the verge of extinction.

The importance of the reserves is growing. They can and must be regarded primarily as unique sanctuaries of a vast collection of natural genetic materials (genetic stock of plants and animals) that nature has created over many hundreds of millions of years. These genetic stocks could not have been preserved in the form of the dead and dried-up materials of botanical herbariums or the zoological collections in museums. They had to be "conserved" alive within the natural, primitive communities in which they had developed. Only in this condition do they constitute invaluable material that can be used for directed selection and hybridization. It is easy to imagine the boundless possibilities for such biological "constructions" if we remember that the basic (species) composition of plants and animals introduced into agriculture is very limited and has not been extended to any appreciable extent for a long time. We should also note the recent productive "explosion" in the formation of cultivated plants on the basis of hydridization between different species.

It is not only in the genotypes of plants and animals that nature is rich. Over the long period of its geologic evolution it has created an endless variety of complex natural systems and processes with truly amazing capacities for adaptation to their environment. Today, therefore, the nature reserves are absolutely indispensable bridgeheads for comprehensive scientific investigations aimed at the purposeful transformation of nature.

A comparative study of the phenomena taking place in the primary natural environment and the modification of those phenomena by economic activity (in agriculture and forestry, for instance) will make possible an analysis of the effects of human activities upon the natural environment. The laws derived from this research can and should be widely used to work out ways and means for secondary purposeful exploitation of nature.

The scientific work on our nature reserves, as it is conducted, does not meet today's requirements. Research should be carried out in a unified program, and the leading nature reserves, at least, should be placed at the disposal of major research institutes.

The preservation of the natural environment and its purposeful improvement is of great importance in meeting cultural, human requirements. An acute and pressing problem in modern society is the provision of recreation in natural conditions for the urban population. This problem has until recently been confined to the establishment or expansion of health zones and resorts. Today, however, the creation of recreational facilities for the urban population outside the places where they permanently reside, calls for specially organized areas combining the original natural conditions, or such conditions artificially modified, with building and facilities to accommodate large numbers of visitors.

Thus the problem of making effective use of the natural environment for rest and recreation is becoming a new branch of scientific investigation, as well as an important part of the economy.

The practical implementation of the proposals in a general plan for the transformation of nature in the Soviet Union confronts Soviet geographers with new and important problems. In order to solve these problems successfully scientists must be given extensive and systematic assistance—their work must be incorporated in the state plan for scientific research, and they must be provided with the necessary personnel and funds.

SUGGESTED ADDITIONAL READING

THE NEW CONSERVATION

THE LAST LANDSCAPE. W. H. Whyte. Doubleday & Co., Inc., Garden City, 1968.

CONSERVATION AND THE GOSPEL OF EFFICIENCY. S. P. Hays. Harvard University Press, Cambridge, 1959.

THE SUBVERSIVE SCIENCE. P. Shepard and D. McKinley, eds. Houghton Miffln Co., Boston, 1968.

THE ECONOMICS OF WILDERNESS

THE POPULATION BOMB. P. R. Ehrlich. Sierra Club-Ballantine, New York, 1968.

A DIFFERENT KIND OF COUNTRY. R. F. Dasmann. The Macmillan Co., New York, 1968.

PROSPERITY AND THE NATIONAL PARKS

DAMS AND WILD RIVERS. L. J. Carter. *Science,* Vol. 158, No. 3798, 1967.

OVER-USE OF THE NATIONAL PARKS. W. A. Johnson. *National Parks Magazine,* October, 1967.

LAKE POWELL: JEWEL OF THE COLORADO. U. S. Department of the Interior, Bureau of Reclamation. U. S. Government Printing Office, Washington, D.C., 1965.

LANDSCAPE ESTHETICS

A SAND COUNTY ALMANAC. A. Leopold. Oxford University Press, Inc., New York, 1966.

WATER. L. B. Leopold, K. S. Davis, and the Editors of *Life*. Time-Life Books, New York, 1966.

DANIEL BOONE IS DEAD

MAN IN THE LANDSCAPE. P. Shepard. Alfred A. Knopf, Inc., New York, 1967.

WILDERNESS AND THE AMERICAN MIND. R. Nash. Yale University Press, New Haven, 1967.

NATURE AND THE AMERICAN. H. Huth. University of California Press, Berkeley, 1957.

PUBLIC PARKS AND PRIVATE LIVES

MAN ALONE: ALIENATION IN MODERN SOCIETY. E. Josephson and M. Josephson, eds. Dell Publishing Co., Inc., New York, 1962.

THE SOCIOLOGY OF GEORG SIMMEL. G. Simmel. Translated by K. H. Wolff. Free Press, Glencoe, 1950.

AEROBICS. K. H. Cooper, M.D. Bantam Books, Inc., New York, 1968.

NATURE AND THE CITY

THE PARK. J. Hope. *Natural History,* August-September, 1967.

THE MYTH OF THE MACHINE. L. Mumford. Harcourt, Brace & World, New York, 1967.

RECREATION NEEDS IN URBAN AREAS. R. C. Weaver. *National Parks Magazine,* December, 1967.

THE MYTH OF THE PEACEFUL ATOM

CHANGE, HOPE AND THE BOMB. D. E. Lilienthal. Princeton University Press, Princeton, 1963.

LIVING WITH THE ATOM. R. Calder. University of Chicago Press, Chicago, 1962.

THE INTEGRITY OF SCIENCE: A REPORT BY THE AAAS COMMITTEE ON SCIENCE IN THE PROMOTION OF HUMAN WELFARE. *American Scientist,* 53, June, 1965.

BANANAS IN VERMONT

SCIENCE AND SURVIVAL. B. Commoner. The Viking Press, Inc., New York, 1967.

MOMENT IN THE SUN. R. and L. T. Rienow. Ballantine Books, Inc., New York, 1967.

THE FRAIL OCEAN. W. Marx. Ballantine Books, Inc., New York, 1967.

CONSERVATION EQUALS SURVIVAL. W. Stegner. *American Heritage,* Vol. XXI, No. 1, December, 1969.

GREEN MOUNTAINS, GREEN MONEY

OUR PLUNDERED PLANET. F. Osborn. Little, Brown and Company, Boston, 1948.

PERILS OF THE PEACEFUL ATOM. E. Hogan and R. Curtis. Doubleday & Company, Inc., Garden City, 1969.

DWINDLING LAKES

WATER QUALITY MANAGEMENT AND LAKE EUTROPHICATION. W. T. Edmundson. *Water Resources Management and Public Policy*. Univ. of Washington Press, Seattle, 1968.

ALGAE AND METROPOLITAN WASTES. R. A. Taft. Technical Report W61-3, Sanitary Engineering Center, Cincinnati, 1961.

INTRODUCTION TO LAKE BIOLOGY AND THE LIMNOPLANKTON. G. E. Hutchinson. Vol. 2 of *A Treatise on Limnology*. John Wiley & Sons, New York, 1967.

ARCTIC WALK

WILDLIFE IN ALASKA. A. S. Leopold and F. F. Darling. Ronald Press, New York, 1953.

TWO IN THE FAR NORTH. M. E. Murie. Alfred A. Knopf, Inc., New York, 1963.

ALASKA: A CHALLENGE IN CONSERVATION. R. A. Cooley. University of Wisconsin Press, Madison, 1966.

THE FLORENCE FLOODS

MAN'S ROLE IN CHANGING THE FACE OF THE EARTH. W. L. Thomas, Jr., ed. 8 vols. University of Chicago Press, Chicago, 1956.

MAN AND NATURE. G. P. Marsh. D. Lowenthal, ed. Harvard University Press, Cambridge, 1965.

FLORENCE: DAYS OF THE FLOOD. F. Nenoini. Stein & Day, New York, 1947.

THE KING BESIEGED

THE POLITICS OF CONSERVATION. M. Frome. *Holiday,* February, 1967.

MINERAL KING GUIDE. P. Adler. La Siesta Press, Glendale, 1963.

THE ECOLOGY OF A RESERVOIR

DEATH OF THE SWEET WATERS. D. Carr. W. W. Norton & Co., New York, 1966.

PRINCIPLES OF MICROBIAL ECOLOGY. T. D. Brock. Prentice-Hall, Inc., Englewood Cliffs, 1966.

ALGAE IN WATER SUPPLIES. C. M. Palmer. Public Health Service Publication No. 657, Department of Health, Education and Welfare, Washington, 1962.

ESCALANTE CANYON

U.S. GEOGRAPHICAL AND GEOLOGICAL SURVEY OF THE ROCKY MOUNTAIN REGION. C. E. Dutton. Government Printing Office. Washington, 1879-80.

STANDING UP COUNTRY: THE CANYON LANDS OF UTAH AND ARIZONA, C. G. Crampton, Alfred A. Knopf, Inc., New York, 1964.

BEYOND THE HUNDREDTH MERIDIAN. W. Stegner. 8 vols. Houghton Mifflin Company, Boston, 1954.

A SONG OF SAN FRANCISCO BAY

SAN FRANCISCO BAY. H. Gilliam. Doubleday & Co., Inc., New York, 1957.

THE RAPE OF THE LAND. B. H. Bagdikian. *The Saturday Evening Post,* June 18, 1966.

THE DESTRUCTION OF SAN FRANCISCO BAY. T. M. Bonnicksen. *Sierra Club Bulletin,* June, 1967.

GEOGRAPHY AND URBAN EVOLUTION IN THE SAN FRANCISCO BAY AREA. J. E. Vance, Jr. Institute of Governmental Studies. University of California, Berkeley, 1964.

THIS WAS SAN FRANCISCO. O. Lewis. David McKay, New York, 1962.

A DETAILED STUDY OF SAN FRANCISCO BAY. San Francisco Bay Conservation and Development Commission, San Francisco, 1966.

A SOVIET PLAN FOR NATURE

PHYSICAL-GEOGRAPHIC ATLAS OF THE WORLD. Academy of Sciences USSR and the Main Administration of Geodesy and Cartography, State Geological Committee USSR, Moscow, 1964. Translation of legend and explanatory text: SOVIET GEOGRAPHY. Vol. VI. No. 5-6, May-June, 1965, American Geographical Society.

NATURAL REGIONS OF THE USSR. L. S. Berg. 8 vols., The Macmillan Co., New York, 1950.

USSR RESOURCES FOR HEAVY INDUSTRY. *Focus,* February, 1969, American Geographical Society.